LOVE AND SACRIFICE
A World War Brings Double Tragedy
to an American Family

Dennis Whitehead

Copyright © 2014 by Dennis Whitehead

All rights reserved. This book or any portion thereof may not be reproduced or used in any manner whatsoever without the express written permission of the publisher except for the use of brief quotations in a book review.

Second Edition, Revised 2017

ISBN-13: 978-0-9863488-5-3

ISBN-10: 0-9863488-5-6

Library of Congress Control Number: 2014919496

MMImedia, LLC
Arlington, Virginia, US

loveandsacrifice.com

This book is dedicated to the men and women of my parents' generation. It is their quiet dedication and sacrifice that laid the groundwork for the riches enjoyed by subsequent generations.

True love is selfless. It is prepared to sacrifice.
- Sadhu Vaswani

Table of Contents

Preface	3
CHAPTER ONE: Norton, Kansas	5
CHAPTER TWO: War on the Horizon	19
CHAPTER THREE: New Life Begins	37
CHAPTER FOUR: The End of the War to End All Wars	47
CHAPTER FIVE: Germany - The Army of Occupation	51
CHAPTER SIX: The Inter-War Years	69
CHAPTER SEVEN: Professor Ollie Reed	83
CHAPTER EIGHT: Back to Benning	91
CHAPTER NINE: Command and General Staff School	99
CHAPTER TEN: Back to School, for Bud's Sake	111
CHAPTER ELEVEN: Duty, Honor, Country - West Point	139
CHAPTER TWELVE: Welcome to the World of Tomorrow	163
CHAPTER THIRTEEN: Sailing to a New Horizon	169
CHAPTER FOURTEEN: The Philippine Islands	173
CHAPTER FIFTEEN: War Closes In	189
CHAPTER SIXTEEN: Laura	195
CHAPTER SEVENTEEN: Starting a New Life	217
CHAPTER EIGHTEEN: Departures	229
CHAPTER NINETEEN: Safe Landings	237
CHAPTER TWENTY: Italy	253
CHAPTER TWENTY-ONE: Normandy	265
CHAPTER TWENTY-TWO: Love and Loss	283
CHAPTER TWENTY-THREE: Epilogue	297
Acknowledgements	305
Endnotes	309
Index	323

Love and Sacrifice

Preface

"Go west, young man," a phrase attributed to newspaper publisher Horace Greeley, summarized America's drive for westward expansion - Manifest Destiny. Telegraph lines followed the advance of the wagon trains, and towns sprouted up along the dusty trails. Progress would not be made sitting still. In the latter part of the nineteenth century, the appeal of a new life brought settlers to homestead the open lands on the windblown Kansas prairie. Where native Arapaho and Cheyenne tribes once hunted buffalo, towns began to grow.

Towns grew with their families, such as those of Ollie Reed and Mildred Boddy. Some stayed to work the land, while others moved on to new chapters in their lives, such as Ollie and Mildred.

From their humble rural Kansas roots, Ollie and Mildred went on to live extraordinary lives with a pioneering and adventurous spirit, all the while holding fast to their simple ethos of dedication to God, country, and family.

Ordinary people have the potential to live exceptional lives, and the Reeds proved this to be true. Whether Mildred was teaching a church group the lessons of the Christmas season through the ornaments on a tree, or Ollie was in charge of a regiment of young men heading into battle, the Reeds readily accepted leading roles without expecting credit or adulation. They were raised in the tradition of making personal sacrifices to benefit others, and to expect nothing in return.

The Reed family - Ollie, Mildred, Ollie Junior, and Theodore, were an exemplary American family during a time when Americans had faith in their political leaders, institutions, and the essential goodness of everyone.

In this book, we are privileged to join their travels through the first half of the twentieth century, guided by the intimate words of Mildred Reed's memoirs, and the personal letters among family members through courtship, love, longing, and loss.

What makes this story truly unique is Mildred Reed's remarkable compilation of historical family materials - letters, photographs, clippings, and artifacts. All of Mildred Reed's collected memories were graciously made available to this author by her grandchildren.

This is a story of dedication, love, and sacrifice; one that blossoms straight from the heart of the American character.

Norton Junior High School 8th grade class, 1911.
Mildred Boddy, seated, second from left and Ollie Reed seated, left on the ground.

CHAPTER ONE

Norton, Kansas

Mildred Boddy, age seven (Courtesy Patricia Tharp)

Youth sped past with the carefree abandon of "the good old days." Hayrack rides, ice cream socials, box suppers, taffy pulls, slumber parties, run-sheep-run on summer evenings, ice skating on the creek in winter. There was a lot going on and I didn't miss much.
—Mildred Boddy Reed

IN 1896, NORTON was a small northwest Kansas town born of hardscrabble pioneers struggling against harsh elements and fickle economic times. They made the best of what they could from what little was given and thanked the Lord for their blessings. The 576,000 acres in the thirty square miles of Norton County are situated about twenty miles south of Nebraska and fifty miles east of Colorado. A 1901 issue of *Western Resources Monthly* magazine described Norton in glowing terms:

> The climate is all that could be asked. The county has an elevation of about 2,300 feet above sea level, and the air is dry and exhilarating, free from malaria of all kinds. The land is very fertile, easily tilled, free from stone, stumps and weeds and very productive.[1]

Closer to reality, Norton's ten thousand inhabitants endured extremes of heat and cold, tornadoes, hailstorms, floods, and infestations. The fortunes of the region's wheat farmers hung on every incoming breeze.

By 1896, the bright lights of the 1893 World's Columbian Exposition in Chicago had not yet reached Norton. Horses and buggies plied the town's one main street under a gas-lit glow, in a year when the thirteen gasoline-powered vehicles were built and sold by the Duryea Motor Wagon Company of Springfield, Massachusetts. Around that same time, Henry Ford was building his first automobile behind his Michigan home. But Norton, remote as it was, remained untouched by these developments for sometime to come.

On August 9, 1896, Mildred Boddy was born into the windblown prairie town of Norton to Della Sarvis Boddy, a twenty-year-old local beauty, and Bert Boddy, a clothing store clerk. Della's mother, Lucinda Biggs Sarvis, was born in Illinois, but her American roots went back to the seventeenth century when Richard Whitaker crossed the Atlantic from England in 1666 to join the Fenwick Colony, a Quaker settlement in New Jersey. In the Civil War, Lucinda's husband, Samuel, fought with the Iowa Volunteers for the Union. In the late 1870s, Samuel, Lucinda, and three of their children, including Della, traveled by wagon train to settle in Norton.

Mildred and Della Boddy, 1897

In 1899, Bert was elected Recorder of Deeds after winning a tied race, according to Mildred's memoirs, by flipping a nickel against Dave Bruner.[2] As the *Norton Courier* reported on November 16, 1899, "The official count discovered a tie between Bruner and Boddy for Register of Deeds and at the casting of lot for the place, Boddy won. It was a republican year, you see."[3]

Barely a month earlier, on July 18, 1896, Ollie William Reed was born to Orville and Mary Plusky Reed. Orville had come to the Norton area in 1878, as a homesteader, staking his claim in nearby Leota Township. He headed west after graduating from Upper Iowa University in Fayette, Iowa.[4] He completed the business program at the college in 1875, known as Hurd's National Business College, offering "all that is taught in the leading Business Colleges of this country at one-half the expense." After completing the six-month business program with a Masters of Accounts certificate, Orville moved into the Normal Department the following year to train as a teacher,[5] putting all of his education aside to seek new opportunities in the soil of a free homestead in Kansas. In 1880, he married Mary Plusky and the couple settled in Orville's Leota, Kansas homestead, bringing a son, William, into the world. In 1886, while Mary was pregnant with another boy, Jesse, the Reed family moved into Norton where Orville took jobs as a mail carrier along Route 5 and a custodian in the courthouse. The growing family needed the more stable income the town could offer. Ollie would be the fifth of eight children in a family with five sons, two older and two younger than Ollie.

The births of Ollie and Mildred marked the humble beginnings of two people whose lives would be woven into the fabric of history, one deeply rooted in the Kansas soil, reaching both triumph and tragedy with quiet grace and humility. The future held unlimited possibilities and as they grew, Ollie and Mildred looked forward to a future together, discovering all that life could provide. Experiencing the world beyond the Kansas prairie was their dream, while family and friends remained at the heart of their lives. Ollie and Mildred dreamed big and lived their lives to the fullest, dedicated to God, country, and family.

As the new century unfolded, signs of the new world gradually began appearing in Norton. In 1902, telephone lines were installed, crisscrossing over the streets. The first telephone operator, known as a "Hello Girl," Miss Clara Stine of Kansas City, was hired to operate the "hello board," connecting Norton residents with one another and the town to the outside world. Upon her arrival, the *Norton County Centennial* noted that Miss Stine was "a bright young lady and experienced." Three years later, new electric lights illuminated downtown Norton. It would be some time until electricity reached the farming communities outside of town.

Kansas was admitted to the Union in 1861 as a progressive, populist, and anti-slavery state. Slavery had been outlawed in the western territories under the Missouri Compromise of 1820, so-called as the exception to the rule was the proposed state of Missouri.

The 1854 Kansas-Nebraska Act reversed this by allowing popularly-mandated slavery in the territories, inciting abolitionists to move west and take a stand. Pre-Civil War skirmishes between the abolitionist Jayhawkers of Kansas and the pro-slavery Bushwhackers from neighboring Missouri were commonplace. Abolitionist John Brown led his supporters against pro-slavery forces in the state, including the 1856 Battle of Black Jack, regarded as the first battle of the Civil War.

Norton County was incorporated soon after Kansas was admitted to the Union, but a tinge of the Old West remained alive in the region. The state song, "Home on the Range," romanticizes the time when buffalo roamed portions of the high plains, as they still did in the early 1870s.[6] Wild Bill Hickok became marshal of Abilene, Kansas, in 1871, and chased outlaw John Wesley Hardin out of the state. In 1892, the Dalton Gang robbed two banks in Coffeyville, resulting in a shootout that killed numerous members of the gang and several townspeople.

A 1903 feud between Chauncey Dewey's Oak Ranch and the sod-busting farm family of Daniel Berry, and his three sons over five dollars owed for a feed tank sold at auction erupted into a gunfight where Daniel Berry and two of his sons were killed, and another wounded in the shootout. When Dewey and his cowboys were found not guilty, the jurors were hung in effigy outside the Norton County Courthouse.[7]

In 1896, a presidential election was underway. On the day Ollie was born, July 18, 1896, Democratic candidate William Jennings Bryan was greeted by ecstatic crowds in his hometown of Lincoln, Nebraska. Bryan was supported by the People's Party, better known as the Populists, and Kansas, a populist state, enthusiastically supported Bryan over his rival and eventual winner, William McKinley.

On August 15, 1896, *Emporia Gazette* editor William Allen White published an essay critical of populism, titled, "What's the Matter With Kansas?":

> Go east and you hear them laugh at Kansas; go west and they sneer at her; go south and they cuss her; go north and they have forgotten her. Go into any crowd of intelligent people gathered anywhere on the globe, and you will find the Kansas man on the defensive. The newspaper columns and magazines once devoted to praise of her, to boastful facts and startling figures concerning

Love and Sacrifice

her resources, are now filled with cartoons, jibes and Pefferian speeches. Kansas just naturally isn't in it. She has traded places with Arkansas and Timbuctoo. What's the matter with Kansas?[8]

By this time, Bryan's Populism had lost steam and his Democratic Party was losing Kansas. In 1896, the state went for Bryan, but in 1900, Kansas helped re-elect McKinley and his vice president, Teddy Roosevelt, who made a whistle stop in Norton during the campaign. With the help of Bryan in the Midwest, Democratic candidate Woodrow Wilson won Kansas in both the 1912 and 1916 elections. During the 1912 campaign, Wilson stopped in Norton to deliver a speech from the back of his campaign train.

In 1913, President Wilson appointed Bryan his Secretary of State, but Bryan resigned in June 1915 to protest Wilson's inching toward war in the wake of the sinking of the RMS Lusitania. Despite stepping down, Bryan remained a Wilson supporter, under the campaign banner of "he kept us out of war."

Politicians were not the only newsmakers in Norton. Documenting the people and events in Norton during the early part of the twentieth century was photographer Charlie Reed, no relation to Ollie. In 1909, he captured the first photograph of a tornado.

Two years later, Reed documented the fatal crash of pioneering aviator John J. Frisbie at the Norton County Fairgrounds. News of the sensational crash made it to the front page of the *New York Times*:

1909 tornado over Norton, Kansas.

Photographs by Charles Reed, courtesy of the Norton County Historical Society.

September 2, 1911

CROWD GOADS AIRMAN TO FLIGHT AND DEATH

J.J. Frisbie Goes up in Crippled Machine

Kansas Spectators Call Him a Faker

Wife Denounces People Who Hooted Him

Amid the commotion and calamity, Mildred and Ollie lived simple childhoods. Ollie was Huck Finn loose, barefoot in the countryside, fishing in Prairie Dog Creek, and hiking through the hills. His future would be on a farm. Mildred was Becky Thatcher, a self-assured young woman who liked wearing pretty dresses with a bow in her hair. She was taught by her mother to be very much her own person.

Ollie Reed, center, in left photo; Mildred Boddy as a child on right.

The idyllic life in small-town America at the turn of the century could also be mixed with personal tragedy. The Reed family never quite recovered from the sudden death of twelve-year-old Hazel on February 6, 1904. She contracted both diphtheria and scarlet fever, and without the antibiotics available today, passed away on that cold winter's day.

Mildred's father, Bert, who had never taken a sick day in his life, fell ill with a fierce headache the day after Christmas in 1903. At the age of thirty-four, Bert suddenly died from a stroke. The morning after Bert's death, Della came into her daughter's bedroom, asking her to say her prayers, "Your Papa is with God this morning."

Mildred, just seven years old, immediately imagined the nice trip her father was on. "I hope he has a lovely time. When will he be home?"

Years later, eighty-three-year-old Charlie Kennedy, a Norton native, remembered Bert in a letter to Mildred:

> In my mind's eye, I can picture him yet as we rested our horses on a hill top out on the prairies, singing one of his favorite songs at the top of his voice (and disturbing no one)… "Oh, Kansas land, sweet

Kansas land, as on highest mount I stand. I look across the fields of grain and wonder why it never rains."

While Mildred's father was healthy until the day of his sudden death, her mother was a frail woman for much of her life. The strain of childbirth taxed her to the point that Bert and Della had only one child; Bert did not want to put his wife through that ordeal again. Della suffered from angina pectoris, a heart condition, and had always been encouraged by her doctor, "to sit on a cushion and sew a fine seam." In spite of her delicate health, she assumed the job of Register of Deeds upon her husband's death, and was subsequently elected treasurer of Norton County, later with Mildred as her assistant.

Della grieved the loss of her husband for years, and dedicated herself to raising her daughter. Norton women whispered about Della, "Widow Boddy is like a marble statue to everyone but that child."

A new Baptist preacher, Thomas Jeffries Duvall, came to Norton in 1914. He caused quite a stir among the town's women, including Della. Mildred later described Pastor Duvall as "a handsome Kentuckian."

"He was forty-five, a widower, a smooth-talker, a 'Southern' gentleman - an exciting change from the western farmer-type men around Norton," Mildred recalled.

All of the eligible women of Norton "set their caps" for Pastor Duvall, but he only had eyes for the attractive and devout widow Boddy, and she prevailed, according to Mildred.

Della Boddy and Reverend T.J. Duvall were married on June 14, 1915, some twelve years after the death of her first husband.

Reverend Duvall was born in Nolan, Kentucky, and graduated from Georgetown College in 1893, and the Louisville Seminary in 1895. He served as pastor at several Louisville area churches before his assignment to Missouri in 1903, moving to Norton in 1914. That same year, he authored the book, *Better than Divorce*. A 1916 history of Norton County noted Reverend Duvall's place in the community:

> [Reverend Duvall] is loved and respected by the good people of the entire community. He is a tireless worker in the field, and a kind and courteous gentleman to meet. Rev. Duvall's father was a captain in the Confederate army but the Reverend is a true union

man through and through in spite of his Southern blood and teaching.[9]

The town of Norton was enamored with Reverend Duvall, but Mildred resented his attention to her mother. Before they were married, she would eavesdrop on their conversations and was repulsed at the sight of them kissing. She wrote of feeling like a baby bird being crowded out of its nest. "There started to grow in my inner consciousness a resilience that I would need later to cope with what Life had in store for me" she wrote in her memoirs.[10]

Mildred chose to spend most of her time at her grandmother's house when the reverend was at home.

Mildred's romantic interests developed as she entered her teenage years, but very slowly - as dictated by her mother. She and Ollie met through the Sunday school class taught by her mother, though it was some time before they were allowed to court. She recounted this in a 1979 letter to her niece, Patricia Tharp:

> Mama taught a high school boys Sunday School class (Baptist) for ten years or so – most unusual at the time. She called it Boddy's Baptist Builders. They had a class meeting once a month at our house. I can remember them going home singing, "Tell Mother I'll Be There in Answer to Her Prayer."
>
> They made over me when I was little. Then, when I became a teenager – was I the envy of all my girl friends! I was not allowed to

Boddy's Baptist Builders - Della, center, and Ollie to her left.

stay in the room for their business meeting, of course, but helped serve refreshments – popcorn balls or doughnuts, or once a year oyster stew! My Ollie was in the class. I was not allowed to go out with boys until I was 16. Already tagged Ollie when the magic hour arrived!

Tall and stocky, Ollie Reed was known for his kind, gentle personality. The inscription next to his senior high school yearbook portrait keenly attested to his true nature:

Obliging best suits this young man,
Doing for others what he can.

Ollie's profile in the 1915 Norton HS yearbook. (Norton County Library)

Ollie loved playing football, his husky stature making him a natural.

"Tubby" is our star tackle,
About this man we have cause to cackle.
On the tackle swing he hits with a "bing,"
And then you should hear the lines crackle.

1915 Norton County High School football team. Ollie is second from the right, middle row.
(Norton Public Library)

Ollie appeared as the lead character in the senior play, acting the role of Lieut. Jack Wilson.

Left: Ollie, as Lieut. Jack Wilson, is seated between Mildred Boddy, as Frank Burton, and Marie Brown, playing Barbara Burton. Above: Lt. Wilson holds onto Doris Meredith, played by Eva Rhodes, against the desires of Italian Count Andreas Cassivilli, portrayed by Ross Hicks, in the senior play.

In their 1915 yearbook "Boneheads" page, Ollie is quoted, saying:

> Ollie, after going through the part of the Senior play where he had to kiss Eva R. suddenly exclaimed: "O, let's do that again."

A futuristic front page from the *Norton Daily Telegraph* displayed in the 1915 yearbook foresaw how life would be ten years into the future, May 1, 1925.

The story featured Sheriff O.W. Reed getting into an argument with the manager of the local barbershop over who had known the most in American History during high school. Mildred Boddy, the make-believe proprietor, editor, and publisher of the *Daily Telegram*, reported the argument under the headline, "Duel in Barber Shop":

> A heated debate ensued ending in a duel of such ferocity that it was necessary to call in the chief of police. Joe Casey, the well-known pugilist who with his accustomed ease and grace, intervened and marched the peace breakers before Judge Pillsbury who collected a large fine from both men to reward Mr. Casey for his services.

In reality, all three boys were good friends, particularly Ollie and Joe Casey, who were football teammates and served together on the student council, Casey as president and Ollie as treasurer. "Judge" Veda Pillsbury was fellow senior, a serious young woman about whom it was written alongside her yearbook picture: "Virtuous and in conduct moral. Quietly evading every quarrel."

That same faux newspaper published a letter to Ollie from Arley Lockard, who left Norton in his freshman year. Writing from London on April 10, 1925, he opens the letter: "Dear Ollie - Well old man, the war is over and the thought fills me with joy."

★★★★

In 1915, Norton was as far from the battlefields of Europe, but there was a sense of America's growing role on the global stage. Automobiles opened new horizons to young people, and their wanderlust steadily grew with the songs about saying hello to Frisco, doing the hula hula in Hawaii, and visiting Holland for tulip time.

"Poor dear Ollie was frustrated," Mildred wrote, "because he couldn't take me away from it all. He was 18, no job, no future without an education. I remember how he bristled when the reverend called me 'girlie.' I didn't like it either but was pleased that Ollie was so protective... Ollie walked me down to Grandma's, where Auntie [Myra Sarvis] and I were to sleep that night. When I went into the house, Auntie tried to tease a little to cheer me up by saying, 'I saw him kiss you.' She really hadn't and was she ever shocked when I said, 'If you had kept looking, you'd have seen him kiss me several times.'"

Mildred hadn't allowed Ollie to kiss her by this point in their courtship, but she enjoyed goading her prudish aunt.

Ollie and Mildred were steady dates in their senior year, but Mildred didn't like being tagged "Ollie's girl." A new boy had come to the school, eliciting the curiosity of all the girls, including Mildred. She figured she had dated Ollie long enough and broke up with him.

To this, Ollie replied: "Alright but listen here - if you ever take me back it's for keeps, you understand? I've had enough of this on-again, off-again business."

Mildred flatly told him, "Don't worry, I won't want you back."

After dating the new boy, whose name was never mentioned, Mildred decided he wasn't as fun as Ollie. She tried everything to get her old beau back, but he didn't respond. Finally, at a Halloween party, boys bobbed for apples to pick their dinner partner. Ollie surfaced dripping wet with the smallest apple in the tub. He told Mildred that he had hoped it might have her name on it, as she was the smallest girl at the party - and it did.

Afterward, he asked Mildred if he could walk her home. At the front door, he asked her why she had allowed him to bring her home.

"Oh, I discovered I liked you better than I thought I did."

With that, Ollie grabbed Mildred and kissed her for the first time. He ran down the street, as Mildred wrote, "clicking his heels."

"In those days a kiss was practically a promise to marry," she recalled. "I had kept a diary for five years but that night I wrote in it for the last time. I figured it was the highest moment in my life, nothing could ever top it - Ollie had kissed me."

Soon after high school graduation, Mildred was stricken with tuberculosis: "I discovered that I wasn't as important in the scheme of things as I had always thought I was. The world rolled right along with me flat on my back." She wrote to her niece Patricia Tharp in 1979, "All my friends went off to college or to teach - and I who had been the sorta hub of activity - left high, dry, and handsome. Learned I wasn't nearly as important as I had thought I was. Valuable lesson for the rest of my life."

Lucinda Sarvis spent long hours with her gravely ill granddaughter, who was bedridden for six months, telling her stories of pioneer life and of their ancestors dating back to 1666. "It made me realize that I was a link in the Continuity of Ages."

One early summer day, a year after high school, Mildred's mother leaned her head back and closed her eyes.

"Oh Mama, don't do that. That's the way I see you in my dreams."

She smiled. "Do you think I'm going to die?"

"Yes."

"I do too," Della replied.

In the evening of June 27, 1916, Mildred was lying in a hammock in the backyard reading Booth Tarkington's *Seventeen* aloud while her mother crocheted lace for pillowcases to add to Mildred's hope chest.

As a circuit preacher, Reverend Duvall would be gone for days at a

Love and Sacrifice

time. At bedtime, they went inside, and her mother immediately fell ill. She went into convulsions and was, as Mildred wrote, in great agony. Della died of heart failure and uremic poisoning the next day.

Mildred cynically prayed, "You are not fair, God, to let her suffer so. She is so good and has always worked hard for you. You've got to help her." Within half an hour, "my precious Mama was dead. That's the last time I've been sassy to the Lord!"

In the safe deposit box, she found a letter written by her mother:

My darling Mildred,

All winter and spring I have had a feeling, foreboding, premonition, or whatever it may be, that I will not live very long. I know of no particular reason why this should be but I cannot get away from it, and in case I should leave you, I know you would wish to know something definite as to my wishes for your plans.

First - my property. I have just made deeds to leave my real estate to you. The property all came to us by the forethought of your father for our welfare, and it is only right that it should all be yours. It would probably be best for you to sell our house so you will have funds for a college education.

If you sell the house and have plenty to do it, I wish you would give Mr. Duvall two hundred dollars, and Grandma, one hundred, as a token of my love for them.

You will have my $500 life insurance, which will more than cover any immediate expenses. My own last ones I would want very simple and inexpensive. I would want to be buried in my wedding dress. I know you will be overwhelmed at first, if my premonition proves true - but your life is all before you and I pray God the future holds much happiness in store for my darling.

Face things as they are, the bright side out always, make the best of everything. Take care of yourself, dear; if you do, I believe you will be well and strong. I do not want you handicapped by poor health as I have always been.

I want you to go to Ottawa College as we have planned, for at least two years; if you wish to make other plans then, I believe you will be old enough to know best what to do.

God has been good to us to spare us to each other so long. Always seek His guidance in all that you do. You have been so precious to me, Dearest, and such a comfort.

Even in the other world, I believe I will not be so very far away from you — and always I will ask Our Father to take care of my precious baby.

With a heart full of love

Your Mother

The Norton newspapers were effusive in their grief over her passing

The news of the sudden death of Della Duvall, county treasurer, was a terrible shock to the community, especially to us who have known her since her earliest girlhood. We met her that evening going home and she appeared in her usual health and cheerfulness. By eleven o'clock there naught but the body in death.

Nature's ways cannot always be fathomed.[11]

Reverend Duvall was upset that Della had not provided for him in her will. Theirs had not been a happy union, lasting only two weeks past their first anniversary. Not long after Della passed away, Duvall handed Mildred a bill for the expenses incurred when he took Della to Kentucky for their honeymoon. After this, Mildred's uncle John Stapp had Reverend Duvall sign a quit-claim deed that he would never ask for money again. In 1918, Reverend Duvall married Cora L. Williams, a forty-five-year-old "wealthy spinster," as Mildred called her, and they were off to his next church in Abilene, Kansas.

With both of her parents gone, Mildred faced an uncertain future.

MILDRED BODDY. C. N.
Mild is her custom—and then,
Wonderful poems fall from her pen.

CHAPTER TWO

War on the Horizon

Company K, SATC.
(Kansas State University Library Archives)

AFTER GRADUATING HIGH SCHOOL, Ollie received news many young men his age would have been thrilled to get. Through the sponsorship of Congressman John Robert Connelly, Ollie was selected for the United States Military Academy at West Point. But, rather than packing his bags and heading east, Ollie spent the summer after his senior year working on a farm and thinking about the course his life would take as a career military man. Was it for him? Mildred later remembered Ollie's tough decision:

> That summer, he told me, "I am giving up my appointment to West Point. I want you to be my wife, and I wouldn't ask you to live the life an army woman has to. She never has a settled home, or garden, or any of the things a woman likes. I'm changing my plans. I will go to Agriculture College and learn about stock. We'll go to South America and raise fancy cattle."

During a visit to Fort Riley over the summer of 1915, Ollie met Cornelia Byram Lewis, the wife of Second Lieutenant John E. Lewis, a 1912 graduate of West Point. Her husband was assigned as an officer in the 10th Cavalry Regiment, best known as the African-American Buffalo Soldiers, serving with General John J. Pershing in pursuit of

Pancho Villa in the Mexican Expedition, Ollie wrote about Cornelia Lewis, "She had been added to the small list of people I believed in. She is the kind of woman I have always thought you would be."

Whatever the trusted Mrs. Lewis told Ollie about the life of a military family was enough to convince him to turn down his appointment and pursue agricultural school.

With nothing holding Mildred to Norton, and estranged from her stepfather Reverend Duvall, she honored her mother's final wish and enrolled at Ottawa University, in Ottawa, Kansas.

"Mama died suddenly when I was 19. I had finished high school at 18 then had rest cure for tuberculosis. Then Mama died. She had plans made for me to go to a Baptist college that fall. I did, but just to kill time until Ollie could support me, Mildred recalled to her niece, Patricia Tharp.

Mildred was one of ninety-six freshmen in a school with two hundred forty-three students, mostly women. After World War I, the student population would double. Located about forty miles southwest of Kansas City, the school was founded by Baptist missionaries as a boarding school for the children of the Ottawa Indian tribe. Congress set aside twenty thousand acres of the Ottawa Indian reservation for the school. In 1865, the name of the school was changed, at the request of the Ottawa tribe, from Roger Williams University to Ottawa University.

Mildred lived in the Charlton Cottage women's dormitory where she played Juliet in a production of *Romeo and Juliet,* opposite her classmate, Helen Park, in the role of Romeo. She and her best friend, Nelle Foree from Tekamah, Nebraska, spent spring Saturdays paddling on the Marais des Cygnes River, careful not to tip over the Otter Dam. She was active in the campus YWCA organization, and was known to have a bit of a sharp wit toward her professors. When, in Freshman Rhetoric, a Professor Ritchie asked: "Why do words have roots?" Mildred chimed in, "So the language can grow."

A little over one hundred miles away, Ollie enrolled in the agriculture program at Kansas State Agricultural College (KSAC) in Manhattan, Kansas. KSAC was the first land-grant college in the

United States, established by Congress during the Civil War in 1862. Land-grant colleges carried a mandate to provide three basic areas of study: agriculture, engineering, and military science.

Ollie joined the Student Army Training Corps (SATC), renamed the Reserve Officer Training Corps (ROTC) in 1920. Marching uniformed students were a frequent sight on campus, as was the sound of small arms fire echoing across the campus green.

SATC cadets march on campus. (Kansas State University Library Archives)

The crisis across the Atlantic, in its second year by the time Ollie entered KSAC, was drawing greater attention in the American consciousness, particularly after the May 1915 sinking of the passenger liner RMS *Lusitania* by a German U-boat. Among the nearly twelve hundred passengers and crew who died in the attack were one hundred and twenty-eight Americans. The public was outraged, and what had been an isolationist sentiment took a decided turn against Germany. Wilson resisted calls for retribution against Germany in the wake of the *Lusitania* sinking, but at the same time called for the mobilization of American armed forces, causing Bryan to resign. Though Wilson won re-election in 1916 under a campaign banner of "he kept US out of the war," it wasn't long before he broke with his slogan.

On March 21, 1916, Wilson's former Secretary of State, William Jennings Bryan, who had resigned over Wilson's response to the *Lusitania* sinking, visited KSAC to speak on the topic "War and Its Consequences to Us." Three thousand students and townspeople jammed into the college auditorium to hear the famous orator.

"The European war has no parallel or precedent in all history," Bryan began. "This is a war that some people want us to go into…This war is not a race war, it is not a religious war …The cause of this war is to be found in a false philosophy. . . We should not get down into the mire, as Europe has done today, to conform to a false standard."

In the audience that day was Ollie Reed, who wrote to Mildred about the great orator:

> Did I tell you that I had the great (?) pleasure of hearing W. J. Bryan speak last Tuesday. Take it from me - I think he insulted the school. He talked on preparedness and had the gall to use sarcasm and 'catch phrases.' The nerve of him sure appealed to me, but not much else. When we sang 'America' he stood up there like a clam wondering how many voters there were in the audience.

As Bryan was making his argument to the students at KSAC against entering the conflict, American troops were mobilizing. This time, they were not preparing for defense of the homeland against invasion. Rather, American troops were training for action on foreign soil.[13] When Germany resumed its submarine warfare against civilian ships and British intelligence uncovered a plot where Germany was offering Mexico the return of territory lost in the Mexican-American War in exchange for an alliance, Wilson was left with little choice.[14]

The U.S. Senate ratified Wilson's declaration of war against Germany on April 4, 1917. The House of Representatives followed two days later. Over the course of the Great War, the Kansas State SATC program provided 431 officers and 1,703 enlisted men to the war effort. Forty-eight K-State enlistees lost their lives in the conflict.[12]

At the outbreak of the war, little had changed in American military organization since the Civil and Spanish-American Wars. With passage of the National Defense Act of 1916, the U.S. moved away from its reliance upon militias and toward a centralized, well-trained National Guard and Reserves in support of a standing Regular Army. The SATCs at land-grant colleges and other schools helped to supplement and supply the armed forces on the eve of war.

Ollie was one of six students living upstairs in Mrs. Light's boarding house at 1231 Vattier Avenue in Manhattan, just across North Manhattan Avenue from the KSAC campus. He had a full academic schedule, plus marches, shooting competitions, baseball, football, and leading Boy Scout Troop 3. As a SATC lieutenant, Ollie commanded Company K, which among its accomplishments had the best shooting percentage in the second year of his command.

But a magnetic pull tugged at Ollie's adventurous side. Seeing a covered paneled wagon adorned with a landscape painting, Ollie recognized it as exactly what Mildred would like for a westward adventure. For such an expedition, he agreed that Mildred could bring whatever she wanted as long as Ollie could pack his two rifles, a shotgun, and his hunting duds. The young couple shared a wanderlust reaching beyond the horizon.

In letters, Ollie and Mildred were pictures of American innocence, reading the same Bible passages every day and sharing their dreams for the future. He encouraged her to pursue her interest in writing, and she responded with elaborate stories of children playing in their yard while Ollie sat in an easy chair smoking a pipe. At the same time, she teased Ollie about the attention other men were showing her back in Norton. One letter, about a young man by the name of Crawford, drove Ollie to distraction, but he calmed himself with a Biblical phrase, "Judge not, lest you be judged."

When she mentioned "going with" a fellow named Davis in another letter, Ollie was quick to reply, "I did want to show just where I stood in case of a second Crawford deal." His next letter finally put a stop to her coquettishness:

> You told me yourself that you were in considerable doubt yourself last summer. I felt that if there were any doubts in your mind at present that you would think considerable about such a letter and you would really know whether you loved me or just thought I was a nice kid etc. I knew that you "thought" that you loved me this winter and last fall. Right now you are standing out stronger than ever - and I was not sure. As for "faith" and "trust" in you Mildred, sweetheart this is a part of my love for you and it will last just as long and that love will last as long as I do, for it is a very good part of me.

After a spate of teasing, the couple returned to their romantic daydreams of Ollie getting through school, and then earning as much as one hundred dollars a month teaching. They would save enough money and hit the road, as Ollie wrote:

> Personally I would like nothing better than to go camping until we found some place we liked, unless you wanted to live in town or city, most of the time. Why good lord we won't either of us be over

24 years old when we start and why not chase around for a year or two. We would both live longer, know more and not miss a year or so. For myself I want to have a year or so of recreation before we settle down for life with a pair of broken down broncho's [sic] and girlie for a wife

Then if we fail, we can grin and act like kids anyway—"Tis better to try and fail than never to try," and as I tell the kids, the victor has never came over the pike that hasn't taken a beating at some time. And that is time. In fact its just about as much fun in trying and getting beat as there is in winning. I can't promise to "keep you in the style etc" but I'll sure keep you.

Meanwhile, Pancho Villa's insurrection in Mexico and incursions across the U.S. border were bringing war closer to home. Villa had invaded Columbus, New Mexico, where the Mexican revolutionary killed eight U.S. soldiers and nine civilians, setting fire to the town. President Wilson reacted quickly by dispatching National Guard troops to the border under the command of Major General Frederick Funston, with General John J. Pershing in pursuit of Villa inside Mexican territory.

After the 1916 school term ended, Ollie was mustered into the 1st Regiment of the Kansas National Guard for deployment into the Mexican Punitive Expedition. The 1st Regiment troops arrived at Eagle Pass, Texas, on July 2, 1916, and marched into the desert to set up camp.

For four months, Ollie served with Company I on the border at Eagle Pass, sleeping in tents and with little to do since Villa was on the run and nowhere near. Ollie patrolled, trained and marched, but his real battle was against boredom and homesickness. The only excitement came in August when a hurricane nearly washed away the encampment.[15]

The Kansas regiments decamped from the Texas border on

September 6, 1916, departing by four "truck trains," a method of troop movement that was innovative for its time. Each of these consisted of thirty-three trucks, traveling caravan-style. The 1st and 2nd Kansas National Guard Regiments arrived at Fort Sam Houston two days later, marking the first time infantry troops had been transported in trucks over a long distance, three hundred twenty-two miles in this case.[16]

Ollie had quickly risen through the ranks on the front line of Texas. A natural soldier, he was rapidly promoted from private, to corporal, to sergeant, and returned to school as a lieutenant in the Reserves. His easygoing Kansan ways, coupled with a knack for leadership and teaching, drew the attention of senior officers. Contrary to his expectations, Ollie found he enjoyed military service and volunteered for the First Officers' Training Camp (FOTC).

Private Ollie Reed

When the U.S. declared war on April 6, 1917, the army was facing a severe shortage of trained officers to lead troops, who, with the passage of the Selective Service Act in May 1917, flooded the ranks. Some five hundred thousand "selectees" needed leaders - and fast.[17] Strategic and tactical planning capabilities were lacking in the existing officer corps, as was actual combat experience. Opposition to a standing army, reinforced by the conviction that America needed only to prepare itself for defense of the homeland, further dampened the supply of qualified officers needed for mobilization. While the draft brought thousands of new men into military service, the officer corps remained static and unprepared.

The officer training program needed a speedy overhaul. Adjutant general Brigadier General Henry P. McCain, great uncle of U.S. Senator John McCain, called for the prompt opening of a nationwide officer training program. At the same time, Secretary of War Newton D. Baker directed construction of sixteen new cantonments, literally military cities, across the country. These cantonments would house each of the new sixteen army divisions of the National Army, numbering seventy-six through ninety-one, added in the American mobilization effort, while training an officer corps sufficient to lead troops into combat. Similar cantonments were built for new reserve divisions.

On April 23, 1917, just two weeks after the declaration of war, General McCain ordered the first camps to be open and ready to receive their first candidates, just one week later.[18] On May 14, 1917, sixteen FOTCs began training 30,000 prospective officers. Of these, "The First Ten Thousand," as the successful officers were called in an April 30, 1917 memo written by the adjutant general, "should be the best the country has." They would join the half million soldiers in the mobilization effort. The commanding general of the Southern Department wrote that his office had been overwhelmed with applications for commissions.[19]

Word went out on April 24 about the officer training programs, and Ollie mailed his application to the Central Department of the U.S. Army in Chicago, accompanied by the recommendations of the Kansas National Guard regimental commanders. The commanding general in Chicago oversaw the opening of six programs in four camps - two at Fort Benjamin Harrison in Indiana, Fort Sheridan in Illinois, Fort Snelling in Minnesota, and Fort Riley in Kansas, for candidates from Colorado, Kansas, Missouri, and Wyoming.

Ollie's acceptance was mailed on May 3, 1917, with orders to report to Fort Riley five days later.

Ollie at Ft. Riley Officer Training Camp

The officer training course centered on building physical stamina and the fundamental skills required of all soldiers: formation drilling, route marching, and basic marksmanship. Candidates without basic military training had to be schooled in the fundamentals of soldiering, bayonet skills, and close-order drilling, as well as signals training, primarily semaphore flag signals. They also dug a lot of trenches. One Leon Springs, Texas FOTC candidate quipped: "When you get discharged from this man's army you can always get a job as a grave digger."[20]

The first month of training was the same for all candidates, and then the men were divided among infantry, cavalry, artillery, and engineering companies. There were fifteen companies in each camp - nine infantry companies, three batteries of artillery, two cavalry companies, and one company of engineers. Two camps had the added specialty of coastal artillery. Candidates were paid $100 per month.

The hastily constructed facilities weren't especially comfortable. The theoretical and tactical parts of the training were often delivered by a training officer reading straight from an army manual, often in hot or freezing-cold, crowded mess halls. The goal of the training program was to simply prepare new officers not to be completely disoriented on the battlefield. But, how could officers without experience in the European trenches convey this? Camps resolved this by constructing simulated battlefields with trenches, barbed wire, and mock-ups of tanks, and other military equipment.

As candidates progressed through the program, they were required to appear before the dreaded Benzine Board, so-called for a post-Civil War panel set up to remove incompetent officers from service.[21] "No inquisition chamber in the Dark Ages ever controlled the destiny of people more completely than did the Benzine Board, as it judged the frailties and capabilities of these aspirants for commission," Officer Training Camp-trained officer Gus Dittmar later wrote. At the program's end, candidates faced additional scrutiny by a graduation board that examined each candidate's suitability for commission.[22]

All of this - training, evaluation, and commission - was accomplished in just three months, leading to the name by which these new officers would be known on the battlefields of Europe - "ninety-day wonders."

Ollie passed his reviews and had a one-week leave before graduation. He telegrammed Mildred in Nebraska where she was visiting a friend.

"I have leave the week of August 19. I hope you will be in Norton." Mildred was certain she knew the meaning of this.

"I bet he will ask me to marry him," she told her friend Nell. "Let's go to Omaha and buy me a wedding dress."

Ollie and Mildred on their engagement.

Once home and with obligations to friends and family fulfilled, Ollie finally had time alone with Mildred.

"Now that I have a job, I can borrow a hundred dollars and get you a ring, or I can borrow a hundred dollars and we can get married," Ollie proposed.

Mildred was quick to respond, "There's a war on. Men are getting scarce. I'll take you now."

A wedding shower was hurriedly organized in Norton. It was a lingerie shower, featuring "teddies, camisoles, combing jacket, blue lisle hose and a gorgeous pink chiffon beaded blouse you could see through. How daring!" Mildred wrote.

After enjoying the spotlight at the shower, Mildred got a ride home from Ollie in his father's Model T. En route, Ollie accused her of being a spoiled, selfish, and egotistical girl.

"Ollie Reed, if you are trying to make me so mad I call off this wedding, you are off your noodle. I know you are scared - so am I - but you'll see. We'll make a success of it and have a good life together."

Ollie and Mildred were married in her uncle Joe Sarvis's grove on the edge of town that Thursday. Although Mildred was uncomfortable with the choice of pastor, the ceremony was officiated by Reverend T.J. Duvall.

The August 24, 1917, *Norton Courier* noted the nuptials:

Boddy-Reed Wedding

At 7:30 o'clock last evening Miss Mildred Boddy was united in marriage to Lieutenant Ollie Reed, Rev. T. J. Duvall officiating. The newly married couple departed on No. 5 for Fort Douglas, Utah, where the groom is to be stationed at present and will make their home there.

Love and Sacrifice

Mr. and Mrs. Reed have the best wishes of the people of Norton and vicinity. Miss Boddy secured an extensive acquaintanceship throughout the county during the time she served as deputy for her mother in the office of the county treasurer, and her unfailing courtesy and consideration won her the sincere friendship of all with whom she came in contact.

Mr. Reed is one of our self-made young men, in the best sense of the word, and possesses the characteristics which will cause him to mount high in whatever capacity he may be employed.

Trying on the Ring

Norton County is proud of this young couple and that each succeeding day may bring to them added joy is the sincere wish of our people.

Mildred later recalled her wedding day:

> The Baptist preacher, my step-father, Thomas Jefferson [sic] Duvall, tied the knot.[23] Mae Sarvis and Art Reed signed as witnesses. Christine's month-old Betty was the only one who cried. Maybe she had a premonition she'd have to live with us someday. We drove Uncle Sam's wife, Nell Sarvis, and children home to Dellvale, then went to Reed's for cake and ice cream. I can still feel the critical stares of solemn-eyed Eva Mae and ten-year-old Harold. I was not the one Ollie's mother had in mind for him. She'd rather he'd have married Maud Deeley. Her mother was wealthy and Maud was a healthy, good-looking girl. I didn't appreciate it at the time, but years later came to see her viewpoint. I wouldn't have wanted my son to marry a puny little individual who'd always had her own way, and had tuberculosis, to boot!
>
> All of our friends were at the railroad depot to give us a send-off.

It was the girls' last and only chance to kiss my Ollie. I didn't care a bit. He belonged to me now.

From Norton, the newlyweds rode the train to Ollie's first assignment as an officer with the 42nd Infantry at Fort Douglas in Utah.[24] They stopped in Twin Lakes, Colorado for a few days of honeymoon at the Hotel Campion, a large country villa built amid the Rockies. They continued on to Salt Lake City where they made their home in The Covey apartments from August 29 to November 10, 1917.

Mildred on the roof of The Covey.

Second Lieutenant Ollie W. Reed

Freshly minted Lieutenant Reed was anxious to join the fight in Europe. Mildred was aware of his drive and supported him, in spite of her natural wariness and angst.

After graduation from the FOTC, Ollie had the choice of joining the regular army, National Guard, or reserves. He chose the regular army.

Ollie thought it was a sure ticket to Europe.

Fort Douglas an internment camp for German-Americans held for interrogation by the U.S. Marshals Service, as well as 320 German naval prisoners of war. Among these was the crew of the German cruiser SMS *Cormoran*. Pursued by Japanese ships in 1914, the *Cormoran* set out from Tsingtao, China, taking refuge in Guam in December of that same year to refuel and take on provisions. The American governor of the island denied their requests. Rather than risk capture by the Japanese, the Germans opted to wait it out in the harbor, but when the U.S declared war against Germany, the safe harbor turned hostile. A firefight broke out between the Americans stationed on Guam and the crew of the *Cormoran*, regarded as the first shots fired by Americans on German forces in the war. When the fighting ceased, the Germans scuttled the ship. Seven German sailors died - two by heart attack and five by drowning. The rest of the crew was taken into custody and transferred to Fort Douglas in June 1917.

Fort Douglas housed approximately forty thousand men of the Regular Army in hundreds of hurriedly-constructed wooden barracks with poor insulation and few amenities.

The newly-constructed facilities covered about two thousand acres with roads, water supplies, sewers, electricity, and facilities. But, with the ongoing recruitment drive and conscription, the fort was already bursting at the seams. By the time Ollie arrived in August, the camp had grown so much that men were sleeping in tents, on the barracks' floors, and in the gymnasium. Three regiments - the 20th, 42nd and 43rd, plus the Utah Field Artillery, were all crowded into the facilities at Fort Douglas.

To stave off attrition, pay for a private in the military had been increased from fifteen to thirty dollars a month in May 1917. Second lieutenants received $141.67 monthly, or $1,700 per year. Officers attended a School of Musketry, where they practiced throwing hand grenades and performed bayonet drills fashioned on British and French combat styles. They dug trenches and hiked over the arid landscape. In September, the combined forces of Fort Douglas held a massive parade in Salt Lake City, drawing more than thirty thousand citizens to cheer them on.

With cold weather coming and not enough barracks to house all of the men, new quarters had to be found for part of the population. On

October 19, 1917, orders arrived for Ollie's 42nd Regiment to transfer to Camp Dodge, Iowa.

On Saturday, November 10, 1917, the men of the 42nd boarded three trains that would take them from the Salt Lake City Union Pacific Depot to their new assignment at Camp Dodge, Iowa, more than a thousand miles away.

Departure day began on a dark note with the suicide of Sergeant Stanislaus Magreta, a member of Company K, Third Battalion from Detroit. As preparations for departure got underway, the nineteen-year-old sergeant, distressed over the recent suicides of his brother and brother-in-law, pressed the muzzle of his .30 caliber Springfield rifle to his chest in the barracks and fired using a coat hanger rig. A native of Russia who had come to the U.S. with his parents when he was eight years old, Magreta had struggled to sustain his family on his monthly sergeant's salary of thirty-eight dollars.[25]

After more than a day of waiting at the station, the 42nd Regiment was finally on their way at 8:35 a.m. the following morning. Ollie penned a letter to Mildred as his train steamed from the mountains and into Price Canyon:

Somewhere east of Helper, UT

Dear Sweet Loveable Wife

Guess I just want to tell you that I love you more every hour, miss you more and think of you more. I love to think of your quizzical laughing sunshining eye, of your golden hair when it's fluffy and sparkling in the sun…sweet woman of mine, I love you, Babe, love you, love you, love you. Tell me that you love me. Think of me Sweetie.

Pardon this writing but they have run me out of my compartment – are playing pitch there – and I am in the smoker writing on a suitcase. Bought a pipe and it has almost cured me of cigarettes.

We are almost 26 hours late.

In the pre-dawn winter darkness the morning after departing Salt

Lake City, the first of three trains carrying the men of the 42nd paused to make repairs along the curved, rugged, and steep single-track approach to Royal Gorge in the Colorado Rockies, just west of the town of Cotopaxi. The second train, consisting of six baggage cars, nine Pullman sleeping cars, six Pullman tourist cars, and a caboose hauled by a locomotive, came to a halt awaiting the first train's restart. Most of the men in both trains were sound asleep. With the gooseneck coupler repaired, the first locomotive very slowly resumed its eastward climb. As the second train prepared to follow, the conductor ordered the flagman to set flares along the curved track to alert the third train where to slow. The engineer on the second train, assured that the signals were in place, so the second train began its slow climb.

The third train, Ollie's Third Battalion carrier, had been held at the Swissvale station until the two trains ahead had passed through the Cotopaxi station. Receiving the go-ahead, the third train resumed its climb. As it picked up speed around a curve, a bright light blinded the engineer. It was the headlight of his own locomotive reflecting in the rear windows of the second train's caboose. Traveling at twenty miles per hour, he only had time to throw the emergency brake and jump from the locomotive before it rammed into the train ahead, splintering the caboose, and sending a baggage car that had been converted into a hospital car into the Pullman sleeper car, where members of the regimental band and headquarters staff were sleeping. The train's flagman, whose job it was to set the warning flares, had barely managed to leap to safety when he saw the third train coming.

Windows shattered and metal screeched. Sleeping men were thrown from their berths into the cold darkness. The Pullman was telescoped to about half its length. Debris was strewn everywhere and the train cars looked as though they had exploded. Lanterns and flashlights glared in the smoke rising in the frigid early morning air. Amid the confusion, members of the 42nd helped free their comrades from the wrecked Pullman cars. The shaken men of the Third Battalion riding in the third train had no idea what had happened, but they quickly swung into action. Medics used what supplies they could find in the now-destroyed makeshift hospital car. The bodies of the dead were removed and placed alongside the track. A soldier ran into the sleeping town of Cotopaxi, about a mile away, calling for help.

PUEBLO CHIEF
PUEBLO, COLO. TUESDAY, NOVEMBER 13, 1917

RCES MEET BOLSHEVI
3 SOLDIERS KILLED AND 15 INJURED IN D. & R. G. WRECK

Three members of the regimental band were killed and sixteen members of the regiment, as well as five employees of the Denver & Rio Grande Railway, were injured.

The bodies of the three victims were transported west to Salida, Colorado later that morning, escorted by three members of the regiment. Salida was intended only to be the transit point for the bodies, but the entire town turned out to honor the soldiers. A battalion from the 42nd Regiment, including twelve members acting as pallbearers, was quickly sent to the town once commanders learned of Salida's desire to honor the deceased. Salida closed for the day. Four thousand people attended an impromptu funeral procession, led by the Salida Municipal Band playing a funeral dirge. City officials and pastors followed the caisson carrying the flower-draped coffins. The Women's Relief Corps, Red Cross, and twelve hundred Salida schoolchildren joined veterans of the Civil War and Spanish-American War in the procession. Captain T.J. Hampson, a Civil War veteran, draped each casket with an American flag, and an enormous flag was draped over the roof of the rail station for the memorial.

At the station, the pastors of the Methodist, Presbyterian, and Catholic churches gave remarks, closing with a patriotic eulogy to the fallen men. A major from the regiment, the ranking officer on-site, was asked to speak but declined, saying that he was not accustomed to public speaking. As the caskets were loaded onto the train, the major, moved by the outpouring of emotion by the local citizenry, turned to address the crowd, declaring that he had never seen a greater display of honor as "was shown by the citizens of this city in their respect to the dead, whom none knew personally but whom all honored as if they

Love and Sacrifice 35

were relatives."[26]

Two of the men killed in the collision, Sergeant Guy B. Alexander, bandmaster of the 42nd Regimental Band, and Sergeant Clayton P. Preston, the band's drum major, had been classmates at Utah Agricultural College in Logan (today, Utah State University), and enlisted in the army together. Four months earlier, Alexander and Preston wed Viola Allen and Marian Smith in a double ceremony. The newlyweds lived in the same Salt Lake City house until the men's departure for Camp Dodge. Alexander and Preston were asleep in adjoining berths in the Pullman sleeper car when the accident occurred. The third victim, Musician Third Class Fred T. Whitehouse, was on guard duty on the rear platform of the second coach when the trains collided.

Rumors circulated that the cause of the accident was sabotage, instigating a secret inquiry by Robert Lee Craft, a white slave officer in the Bureau of Investigation office in Pueblo, Colorado, but there was not enough evidence to give credence to the speculation.[27]

Colonel Clarence E. Dentler, who had taken command of the 42nd Regiment only days before departure, had ordered one of his officers to accompany the flagman to gather evidence for an army investigation. Removal of the evidentiary materials, as well as the dead and wounded, made subsequent railroad investigations difficult. Despite these challenges, the investigation by the Denver & Rio Grande Railway found the second train's flagman to have been at fault: "This accident was caused by the failure of Flagman Lewis, of train 2d No. 16, to properly to protect the rear of his train."[28] Though not arrested, Flagman Lewis was fired.

Fred Whitehouse's body and the injured soldiers arrived in Salt Lake City, while the bodies of Sergeants Alexander and Preston continued on to Logan, Utah, where two thousand people met the train and escorted them to their homes before funeral services at the Logan tabernacle of the Church of Jesus Christ of Latter-day Saints with full military honors.[29] Joseph F. Smith, the sixth president of the LDS church, asked Senator Reed Smoot to speak at the funeral the next day.[30]

Ollie tried to get ahead of the news before it reached Mildred, telling her about a freight train accident, but word traveled faster than he imagined. Mildred posted a tongue-in-cheek letter to Ollie at Camp Dodge only four days after the accident:

> Say, I am pretty blank mad at you. I don't think you are one bit nice so I don't! and I never will trust you again - so there! You told me it was a freight train wrecked ahead of your train. Of course I had heard about it and when folk ask me if it was your train or a soldier train I denied it heartily and said my man always told me the truth he would know I would have sense enuff not to worry - etc etc etc, but when Mildred Harmonson said Roe wrote her about it busting windows in their train and everything - well I was so mad at you I bawled! I don't think it was one bit smart, young man and after us always agreeing to tell each other everything too.
>
> Well, I'll get even. Maybe I'm awful sick right now - just about to die - but I'm not telling you about it, am I - and I'm not going to either!
>
> So there! Put that in your old dirty stinking pipe and smoke it! Maybe you don't think that's sweet of me to talk that way but I don't care. You don't deserve a nice wife if you don't tell her true things!

With that off her chest, in the next paragraph, Mildred turned to news that Fae Phinney had left school, walking through the mud and rain to Mildred's house to announce that she was leaving home.

CHAPTER THREE

New Life Begins

1917 booklet about Camp Dodge.
(Iowa Gold Star Military Museum)

DELAYED BY THE ROCKY MOUNTAIN CALAMITY, the 42nd Regiment finally arrived at Camp Dodge, located outside of Des Moines. Established in 1909 as a small training site for the Iowa Militia, Camp Dodge rapidly expanded from seventy-eight acres to 570 acres in 1917 to facilitate its role as one of the sixteen mobilization training camps and headquarters for the 27,000 men of the army's 88th Division. Thirty new two-story barracks were built at Camp Dodge and opened with the arrival of the 42nd in November 1917.

The expansion of Camp Dodge cost nearly 5.5 million dollars, over 108 million today. In total, the federal government spent nearly 193 million dollars on construction, today, nearly four trillion dollars.

At Camp Dodge, the men of the 42nd Regiment participated in training programs for small arms, field artillery, gas warfare and intelligence techniques.

With Ollie safely situated and Mildred living in Norton, Ollie searched for a farmhouse where they could live close to the camp. The newlyweds longed for each other, but Mildred was less than excited about the prospect of moving into officers' quarters on the base.

If nothing else, they planned for Mildred to visit Camp Dodge by train as often as possible, and Ollie laid down a set of "travel orders" from his, "Hdqs. Reeds Corp (comprised only of Mrs. O.W. Reed)." The first order "Commander" Reed issued was that "the Reed army will always take a Standard sleeper while en route from one point to another," explaining that "a tourist sleeper is the camping ground of all undesirables who have a dollar bill."

Mildred replied, a little baffled by her sudden immersion in military life:

> I wish I knew more about "army" than I do. Every place I visit the men of the family pounce on me and talk "war'" and "army" and "what does Ollie say and think." I have to try to answer all sorts of questions – how many men in a reg. – how many in a company - how many co's in a reg. – all about the insignia and ranks of different officers. I usually tell them that we don't discuss such things in our family then make a guess at some sort of an answer so they don't think I'm "plum" ignorant.

Hastily constructed housing left soldiers exposed to the cold-weather threat of pneumonia, an ailment that bedeviled military installations nationwide. During the brutal winter of 1917–18, Ollie contracted the disease soon after Mildred joined him, living in a farmhouse attic two miles from Camp Dodge. Thinking that her husband had come down with a cold, Mildred employed her grandmother's remedy, or so she thought - kerosene and sugar. Her patient became violently ill but survived the cure. She later found it was her grandmother's curative recipe for croup (cough), not a cold.

In the spring of 1918, Mildred presented Ollie with the news that she was pregnant. Thrilled about the coming addition to their new family, the young couple again faced separation as Ollie's regiment was ordered east on a temporary assignment to the Picatinny Arsenal in New Jersey, where there were no facilities for spouses. Plus, Ollie had no idea how long he would be stationed there. The Picatinny Arsenal was ill-suited for infantry troops, as the mission of Picatinny was the production and storage of propellant powders and explosives. But, they did have plenty of room to help offset overcrowding at other training facilities. Mildred and Ollie were acutely aware that this assignment meant that Ollie could quickly be sent to France without her seeing him beforehand. Anxious, Mildred returned to Norton in March with Ollie before he headed east. There, Mildred was able to be with her beloved grandmother, Lucinda Biggs Sarvis, when she passed away on April 17, 1918, at the age of seventy-two.

In July 1918, the 42nd moved again, this time north to Camp Devens, Massachusetts. Certain that this would be Ollie's last stop on U.S. soil before entering the war, Mildred moved east from Norton to Manchester, Connecticut, less than 100 miles from Camp Devens. Christine Sarvis, the wife of Mildred's uncle Joe Sarvis, was visiting her family in Manchester. Christine's sister, Nellie Hollister, was a maternity nurse who opened her home at 28 Marble Street to Mildred. Serving as a midwife in her own home, Nellie would see Mildred through her pregnancy and delivery. Mildred could only hope that Ollie would be on hand for the birth of their first child.

Ollie and Mildred in Manchester, CT

Manchester, located on the outskirts of Hartford, was a mill town hosting the world's largest silk mill, Cheney Brothers. Immediately, Mildred was enveloped in the tight community. Across the street, at 29 Marble, lived Esther Rockwell, a Cheney Brother employee, along with her brother Ralph, a carpenter. Sisters Olive and Marjorie McMenenmy, also residents of Marble Street, were part of Mildred's circle of friends.

Camp Devens, another of the sixteen military cities built in 1917, was located on five thousand acres of land leased and then later purchased near the town of Ayer, about forty miles west of Boston. Some of the camp's acreage had been farmland along the Nashua River and some was "sprout" land where trees had been cut for lumber to build the new facilities, leaving only stumps.

Built to house 35,000 soldiers, Devens was overflowing with 45,000 men when the 42nd arrived. The 42nd, along with the 36th Infantry

Camp Devens personnel, December 3, 1918; Ollie's 42nd Regiment is on the far right. (Fort Devens Museum)

Regiment, was assigned to the 12th Division, commanded by Major General Henry P. McCain, the driving force behind the development of the new officer corps in his pre-war role as adjutant general. More than one hundred thousand soldiers of the 12th and 76th Divisions passed through Camp Devens during World War I.

In August 1918, doctors in Boston dealt with their first cases of influenza among naval personnel in Boston Harbor. Soon, soldiers at Camp Devens were falling ill. On September 7, 1918, the day after the Boston Daily Globe published a story saying the outbreak among sailors might spread, a member of Company B of the 42nd Regiment was the first to report to the hospital complaining of a fever, headache and weakness. Regarded as the first case reported of the epidemic by the Surgeon General Merritte W. Ireland in his 1919 annual report, the soldier was initially diagnosed with cerebrospinal meningitis. The next day, a dozen more members of the 42nd reported on sick call. Doctors suspected an outbreak of grippe, the old-fashioned word for the common flu. Quickly, the patients developed severe upper respiratory symptoms, ruling out meningitis and grippe. Perhaps, they thought, it was pneumonia, a persistent problem at over-crowded military bases around the country. By September 10, 142 men had reported ill. Two days later, doctors officially reached a diagnosis of influenza but one they had not seen before. Regardless, Rear Admiral Spencer S. Wood of the 1st Naval District urged people not to be alarmed, that it was simply grippe. One week after the diagnosis of influenza, nearly twelve hundred men were hospitalized. By mid-September, nearly 7,000 men suffered flu symptoms and by the end of October, more than 17,000 men were taken out of action by influenza and pneumonia. There were 688 deaths in September alone, 93 of them on September 26. Another 300 would succumb before the epidemic ran its course in 1919.[31] No antibiotics or antiviral treatments were available to cure either influenza or pneumonia. Doctors and nurses, too, were falling ill and dying from their contact with the infirmed, their risk of exposure heightened by the stress, long hours, and exhaustion.

Camp Devens became a nightmarish scene. Lines of sick men clutching blankets stood outside the hospital in the cold and rain. The lightly insulated wooden hospital units, built to handle a maximum of 2,000 patients, overflowed into hallways and onto porches, with nearly 8,000 men in need of hospitalization. Unlike a typical influenza where a sufferer would spend several days in discomfort and then several more in recovery, this influenza continued to worsen. The ultimate cause of death in most cases was pneumonia that developed from the upper respiratory complications, causing a patient to take on a blue skin color due to the deprivation of oxygen to the blood from a condition known as cyanosis. Patients' skin took on the color of slate and autopsies revealed blue, liquid-filled lungs, typical of pneumonic plague. Some feared the return of the "black death."[32]

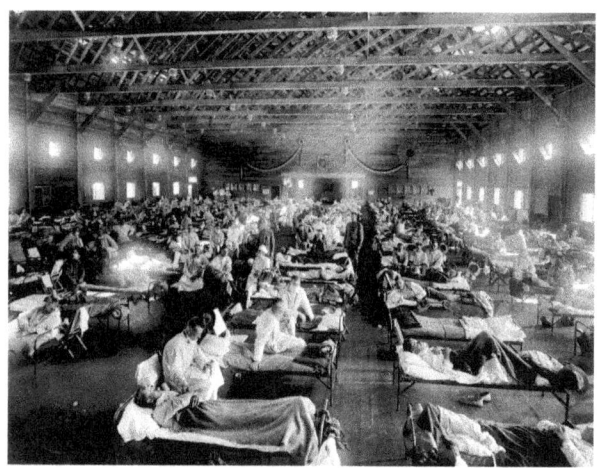

Influenza ward, Camp Funston, 1918
(National Museum of Health and Medicine)

On September 29, 1918, Dr. Roy N. Grist, an Army physician at Camp Devens, wrote to a friend about the horrors he was witnessing:

> This epidemic started about four weeks ago, and has developed so rapidly that the camp is demoralized and all ordinary work is held up till it has passed. These men start with what appears to be an attack of la grippe or influenza, and when brought to the hospital they very rapidly develop the most viscous type of pneumonia that has ever been seen. Two hours after admission they have the

mahogany spots over the cheek bones, and a few hours later you can begin to see the cyanosis extending from their ears and spreading all over the face, until it is hard to distinguish the coloured men from the white. It is only a matter of a few hours then until death comes, and it is simply a struggle for air until they suffocate. It is horrible. We have been averaging about 100 deaths per day, and still keeping it up.

So I don't know what will happen to me at the end of this. We have lost an outrageous number of nurses and doctors, and the little town of Ayer is a sight. It takes special trains to carry away the dead. For several days there were no coffins and the bodies piled up something fierce.[33]

Over the course of World War I, nearly eight hundred soldiers died of influenza at Camp Devens, with more than 13,700 falling ill.[34]

Influenza typically leads to death in the very young and very old populations, measured on graphs in a U-shaped curve. However, in the case of the pandemic of 1918-19, the curve took on the figure of the letter W, afflicting the young and old, but also those ages twenty to forty, men of military age.

When General McCain took command of the 12th Division on August 20, 1918, he promised that the men would be trained and ready for combat in fourteen weeks. Instead, nearly twenty percent of Camp Devens personnel were ill and seventy-five percent of those were hospitalized. The men who did manage to reach France brought the influenza plague with them into the crowded and unsanitary trenches.

At the same time Camp Devens was under attack by influenza in September 1918, Allied forces were launching the Meuse-Argonne Offensive, the effort that would bring the Germans to surrender. But, the spread of influenza, without regard to allegiance or the no-man's land between forces, was just as responsible in bringing the German forces to their knees, as were their combat losses.[35]

Pregnant and afraid for the fate of her unborn infant in a world of disease and death, Mildred relied upon her dreams and imagination for a better world. She wrote to Ollie of one such vivid dream, an idyllic homestead filled with children and joy.

My Husband

I am dreaming tonight of the happy days to come when the war will be over and my man will come marching back again to love and home and me.

I see a vine-covered cottage among the trees. A gravel path leads up to the steps of a broad veranda where on summer afternoons and your days duties are over you will find your wife waiting for you. There will be comfortable wicker rockers and you will rest in the shady coolness and sip lemonade or grapejuice as you read the evening paper. And sometimes will have our evening meal served out there or in the arbor at the side of the house and there'll always be tomatoes and lettuce salad and things my hubby likes to eat.

Then we'll go for a stroll in the garden, won't we, stopping now and then to pull a weed or train a new tendril of a vine or see if any more "everbearing" strawberries are ripe. After a while we'll come back and sit on the steps and watch our kiddies chase fireflies in the gathering dusk. Can't you hear their happy laughter and the joyous barks of the pup that leaps beside them? And sometimes we'll go for a boat ride or to the woods for a picnic or we'll play we're fish and take our family swimming! And toward the end of the summer you'll come home to find the house smelling all spicey and nice and I'll be making watermelon pickles and preserves and jelly and all sorts of good things so our cellar will be filled for winter.

Then autumn days will come – the trees will turn to golden fires – and we'll gather the apples and fruit; and take hikes into the woods nut hunting. It'll be time for Buddy and the rest to be starting off to school again and we'll stand in the door waving to the sturdy little figures laden with books and lunch, until they are out of sight.

Evenings we'll all rake up the fallen leaves and have a big bonfire. Did you love a bonfire when you were a kid? I did. And after a while, winter will be here, and the children will come home from school all rosy and cold – starving for some of the fresh cookies that they'll smell – I expect you'll want some, too, won't you? And the boys will get in some wood and build a fire in the fireplace and after a lot of steaming oyster soup, we'll all gather around the living room fire to spend the long winter evenings – you'll probably be reading and maybe smoking not because you like to, of course, but "to kill the bugs that might otherwise eat up mother's plants." (see how pretty my green ferns and red geraniums look in the big window over there – looking out on a snow-blanketed world?) and Buddy and the other kids will be studying around the library table or playing checkers or dominoes and the little ones will be "building blocks" on the floor, and I'll be rocking the baby to sleep or doing bits of sewing for there will always be socks to darn and tiny aprons and shirts to makes for wee boys and girls.

Bed-time will come soon and they'll troop off to undress and come back in their white pajamas for a good night kiss and to see if they can't beg a story. Then I'll have to go "tuck 'em in" and we can have a little chat all by ourselves, undisturbed except perhaps by occasional "Mama I want a drink" – "Father, when you comin' to bed" – "Ma, Bob and Billie are having a pillow fight," etc. And the wind will whistle around the corner of the house – Woooooo! And oh! Won't I be glad I'll have a nice big warm man to cuddle up to! On Sundays we'll let the kids make candy and crack nuts and pop corn and roast apples, and we'll play the piano and have lots of music – and we'll have oatmeal and waffles and omlete for Sunday breakfasts – and we'll all go to church to-gether (just about be a congregation ourselves won't we!) and the kids will fight to see who gets to sit by "Father"! And maybe we'll go bob-sled riding sometimes and skating. And Thanksgiving we'll have a regular old-time celebration with roast turkey and cranberry sauce and doughnuts and cider and pumpkin pie and everything and we'll have company, maybe – people that we like, not duty friends, and have a hilarious old time. And Christmas, too. We'll decorate the house in pine and cedar boughs and holly and mistletoe and the kiddies will hang up their stockings. Won't it be fun to fill them and to trim the little tree? And it'll not seem like we've been in bed an hour until they wake us up declaring its morning and the want to see if Santa has come. Won't their eyes "bug" when they see the tree all lit up and the things they've wanted most in their stockings? And some years you must be Santa and come jiggling in with your pack! And there'll be goose for Christmas dinner – and plum pudding! And in the afternoon I expect you and the boys will want to go hunting or out to try coasting in their new sleds, and the rest of us will go out in the yard and build a snow man!

Then will come Buddy's birthday – all of our birthdays must be celebrated with a cake and candles if nothing more and sometimes we'll have little parties and surprises for them.

Finally spring will come, then we'll have to get busy! It'll be jolly watching for the birds to come and the first leaves and flowers to open, won't it! And we'll all have our gardens to plan and plant – can't you just smell the freshly tuned earth that we'll be digging in? And see the long slim slick wiggly worms we'll find! Buddy will get a tomato can for their preservation and first thing we know I bet the "women-folks" will be left to do the work 'cause the men of the family will all be off fishing!! And we'll have to set the hens - won't my babies be crazy over the wee yellow chicks? – and prune the hedge and trees (or do we do that in the fall?) and <u>houseclean</u>! "Get out of my way – this is my busy day!" – beating rugs, washing windows – you'll all just have to clear out or else pitch in and help!!! And the boys will leave their shoes and stockings off and begin playing hookey from school. Maybe a circus will come to town and we'll have to take our tribe – little ones at least – the biggest can carry water for the elephants and work their ways in! Then we'll have 'circuses' in our barn for weeks to come and I suppose I'll never be able to locate a pin, no matter if I leave my tray full. And sometimes we'll have little honeymoons off by ourselves, won't we, Lover, when we'll relive the happy days of our first year of married life and pretend we're young again – (that is: if we can help from worrying if the children have stuck all the beans in the house up their noses or drank the fly poison or eaten the matches!).

Do you think I'm crazy, dear, dreaming away the night this way – but it is nice to think about, don't you think so? I haven't included measles and broken bones and other inevitable things in my dreams and of course there must be cloudy times for us, but we'll be happy no matter what happens won't we, My Own, even though things won't be so ideal as I have been picturing them. So you see you've just <u>got</u> to come back to me! You will won't you, my darling? And then we can build some more air castles and maybe work some of them out to-gether.

Remember that little air castle poem?:

> "I would rather be a builder of castles made of air
> To be rebuilded every day, and dwell in fancy there
> With everything to make me glad, the doors all closed to gloom
> And the sunlight of to-morrow shining into every room,
> Than ever keep within the walls of sad things, past or now,
> For tho my castles do not last, they're cheering anyhow.
> And so I build and build again, rebuild from day to day
> And sometime the Master Builder may let my castles stay.

Good Night, O Best Beloved.

Your Wife

CHAPTER FOUR

The End of the War to End All Wars

September 1919, searching for her home in Amiens, France. (National Archives)

THE LONG, BLOODY CONFLICT, known as the Great War, finally ended with the signing of the armistice on the eleventh hour of the eleventh day of the eleventh month of 1918. The influenza pandemic was waning and the worry of Ollie being sent overseas to fight in a foreign war no longer troubled Mildred's mind. She was seven months pregnant and needed some measure of comfort. That her husband would not be thrown into the fire of war was certainly a relief.

An estimated ten million soldiers and seven million civilians lost their lives in the war. Of the 4.7 million members of the military during the conflict, 53,402 Americans died in combat zones and another 204,002 were wounded.[36] The worldwide numbers of deaths due to influenza dwarfed the number of combat casualties. It is estimated that somewhere between thirty and fifty million people died in the pandemic, with an estimated 675,000 Americans among them.[37]

Although Mildred was eight months pregnant in December of 1918, army life demanded that Ollie move again, this time to Camp Upton on Long Island. Camp Devens was overflowing with troops returning from Europe, making space a premium. The same was true at Camp Upton, but the 42nd was moved nonetheless.

Back in the summer of 1915, Ollie had written to Mildred about turning down his appointment to West Point, as he did not want her "to live the life an army woman has to. She never has a settled home, or garden, or any of the things a woman likes." In two years of marriage, Ollie had been stationed at five different posts across the country: Fort Douglas, Utah in August 1917 where his 42nd Regiment was squeezed out after less than three months. Next, it was Camp Dodge, Iowa for four months until tight quarters forced another move to temporary quarters (four more months) at the Picatinny Arsenal. Ollie was then relocated to Camp Devens, Massachusetts in July 1918, only to be crowded out again in December. Now, on the eve of his first child's birth, Ollie was off to Camp Upton, New York, near the town of Yaphank on Long Island, today the site of the Brookhaven National Laboratory. This was the way of army life for the young lieutenant and his wife, and a child would soon be joining the whirlwind.

On January 15, 1919, Ollie William Reed, Junior was born in Nellie Hollister's Manchester, Connecticut home. In her memoirs, Mildred described the origin of baby Ollie's nickname — "Buddy":

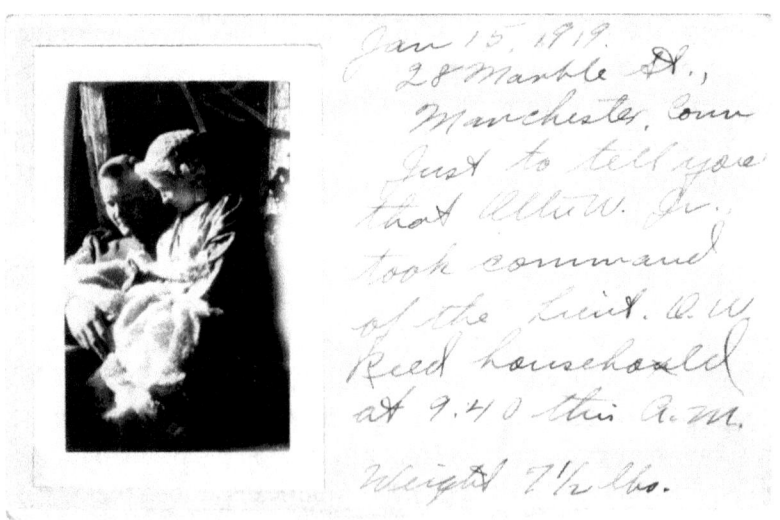

The fall before he was born, we read a clever book giving definitions of Army slang. "Buddy" was a new word meaning "closest friend." I figured the little fellow under my heart couldn't be any closer and I knew he'd be a friend, so he became "Buddy."

The first thing I heard after the birth as I drifted back to consciousness was Aunt Nellie Hollister, who mid-wifed the procedure, saying: "It's your Buddy all right, Mildred."

Having lived in so many influenza hotspots, it was remarkable that Mildred and Ollie, with the exception of his bout with pneumonia, remained healthy. Millions of people fell ill around the globe, so tragedy was never far from home. This was the case with Mildred's aunt, Christine Sarvis, who brought Mildred into Nellie Hollister's home. Christine returned to Norton, her husband, Joe, and two daughters, Louise and Betty, just before Buddy's birth and, on New Year's Day, 1919, succumbed to influenza at the age of forty-two.

Christine Sarvis, left, and Mildred Reed.

Ollie, Mildred and baby Buddy

Traveling the one hundred miles separating the young couple between Ollie's post at Camp Devens, and his wife and child in Manchester was difficult, but the journey from the eastern third of Long Island took time and patience. In good weather, Ollie rode the Port Jefferson branch of the Long Island Railroad, just outside the rifle range on the northern end of the camp, to the ferry for Bridgeport, Connecticut. From there, he transferred to the New Haven Railroad and the fifty-four-mile ride to Hartford, and another ten miles on the New Haven Highland Line to Manchester. When the Long Island Sound froze, Ollie had to travel into New York City to reach Manchester. He would spend more than a half day, each way, on travel, but Ollie didn't care - the journey was well worth it.

In war's aftermath, Camp Upton housed nearly forty thousand men as the headquarters of the 77th Division – the New York City Division, a pretty rough and tumble bunch. An officer later described the recruits as a motley crew made up of "the gunman and the gangster, the student and the clerk, the laborer, the loafer, the daily plodder, the lawyer, men of muscle and men of brain."[38] But as motley as this crew may have been, the war also forged some unique alliances. During the war, members of New York City's street gangs came in as draftees to Camp Upton, declaring a truce to focus their fight on the Germans. The gang from Hell's Kitchen, from the West Side of Manhattan, formed a truce with the Gas House Gang from the East Side to "cook the Kaiser's goose."[39] Songwriter Irving Berlin composed, "Oh How I Hate to Get Up in the Morning" during his time at Camp Upton.

The 77th deployed to France in March 1918, and over six months of combat the division lost 9,294 men and 317 officers. The 77th was also known for its "Lost Battalion," some 554 men from nine companies who were surrounded by German forces in the Argonne Forest. The exact numbers of casualties and survivors has never accurately been determined but the unit, no doubt, lost most of its men to death and capture. They were coming home, and Camp Upton was their demobilization station before returning to civilian life.

Because of its proximity to New York City, Camp Upton was a primary debarkation camp where thousands of soldiers poured in from overseas duty for a ten-day demobilization process and discharge from the army. Soldiers from New York state, Connecticut and Rhode Island were processed at Upton. With their paperwork in order, soldiers were paid and provided transportation to their civilian home.

Ollie's 42nd Regiment was demobilized in 1919 and, on August 28, he was re-assigned to the 50th Infantry Regiment at Camp Dix, New Jersey. Seeing the men returning from combat was deeply disheartening to Ollie as he had not been with them. Indeed, he had been spared the dangers of warfare at a time when he was about to be a father but he was trained to lead infantrymen in combat.

Two months later, Ollie would finally see Europe when the 50th Regiment set sail across the Atlantic as part of the Army of Occupation in Germany. Mildred and the baby would have to wait until the army approved spouses and children joining their soldiers. Again, the Reed family would be splintered and no one was certain for how long.

CHAPTER FIVE

Germany - The Army of Occupation

American troops march in Coblentz, Germany, undated.

AMERICANS ARE FAMILIAR with our ongoing military presence in post–World War II Germany, but little is known about the American forces stationed in the Rhineland region of western Germany in the wake of the First World War.

Over the course of World War I, the empires of Germany, Austro-Hungary, Ottoman Turkey, and the Russian czar had all collapsed. The same Communist revolution that swept over Russia in 1917 raged in Germany. Bavaria renamed itself the Bavarian Soviet Republic (*Bayerische Räterepublik*) from 1918 into 1919. The threads that held Europe together before the war were now all but gone.

The Armistice bringing an end to World War I demanded the immediate evacuation of German forces from France, Belgium and Luxembourg. Allied forces would enforce the demobilization and monitor the movement of German soldiers across the border. Then, American, Belgian, British and French troops would occupy the Rhineland region along the western frontier of Germany to keep the peace and, later, enforce the terms of the Treaty of Versailles. Additionally, the U.S. forces provided a critical buffer between France and Germany.

Two weeks after the Armistice was signed, American troops of the Third Army, led by General John J. Pershing, marched into Coblenz, Germany, where they established their headquarters.[40] On July 3, 1919, the wartime American Expeditionary Forces (AEF) ceased to exist, becoming the peacetime American Forces in Germany (AFG).

As the first anniversary of the war's end approached, First Lieutenant Ollie Reed of Company L, 50th Infantry Regiment, sailed from the port of Hoboken, New Jersey on October 16, 1919, and landed in Brest, France on November 1. Four days later, the 50th marched into Coblenz for duty as part of the Second Brigade, headquartered in Mayen, near Laacher See, a volcanic lake in the Rhineland region of western Germany.

Condition V of the armistice stated, "The areas of the left bank of the Rhine shall be administered by the local authorities, under the control of the occupation troops of the Allies and the United States Armies of Occupation."[41] Initially, the primary task of the AFG was the collection of German military equipment and munitions under the terms of the armistice. Over time, the forces kept watch over the peace along the Rhine River. The U.S., along with Belgium, Great Britain and France, established military governments and restored normal commerce and life along the Rhine, the primary link between northern and southern Europe, as well as a border between Germany and the countries to its west.

The American occupation forces were responsible for general supply and transport to American military personnel and their families, as well as for German civilians, and Ollie's first assignment was as a Rail Transportation Officer with the Quartermaster Corps. As a first lieutenant, Ollie was second in command of Company L. The company commander was First Lieutenant Eugene E. Pratt, a lawyer from Ogden, Utah, and just behind Ollie was Second Lieutenant Charles Q. Lifsey, a 1918 graduate of West Point. In September 1921, Lifsey was selected to represent the 50th Regiment as a member of a ceremonial battalion under the command of General Pershing, for ceremonies in London and Paris, decorating graves of the unknown soldiers in those cities with Congressional Medals of Honor.[42]

Ollie took charge of rail transportation at a time of recurring floods along the Rhine and Moselle Rivers, followed by record-low water levels, both events prohibiting passage of supply barges. The Rhine River, coursing through its namesake Rhineland, was the main supply vein for American forces. Coblenz, the occupation headquarters, had been a supply hub for the German Army during the war, so the transition to peacetime use occurred rather smoothly.

Undated photo from the Reed family collection, Ollie on far left.

With fluctuating river levels, the rail system played an integral role in transportation but American operations had to be resourceful as the lines running through France had either been destroyed by retreating armies, not maintained over the four years of war, or were simply worn out from the tonnage moved in the conflict. Locomotives and train cars were broken down, and there was a shortage of skilled personnel to repair them. Additionally, the French government asked the AFG to find alternate routes, as they needed the rails to help rebuild their war-torn country. In response, American forces opted to utilize the ports at Antwerp, Belgium, and Rotterdam in the Netherlands, transporting goods on rail lines from those ports to Coblenz when travel along the Rhine was inhibited.[43]

Troops and their families needed to be fed and clothed. Existing German stocks and production could not keep pace with even the needs of the local population, let alone an occupying force of nearly equal size. So, importing goods was an essential role of the Quartermaster Corps in the occupation for all concerned. Germany had suffered mightily from four years of blockades and idled civilian production, let alone the loss of manpower to death and disability in war. What little could be grown and found in storehouses was barely adequate to feed and clothe German civilians. The undertaking of the Quartermaster Corps was significant to soldier and civilian, alike.

Quartermaster Corps football team, Ollie second from left, top row.

The tour in Germany was not all work. Ollie, of course, played football as a member of the Quartermaster Corps team in Coblenz, and was involved with the boxing program, including the championship held in the main square of town. With little else to do and unlimited ammunition available, as well as access to training areas, troops constantly trained, making them the most polished soldiers in the army at that time.[44]

On March 23, 1920, the commanding general of the AFG, Major General Henry T. Allen, issued permission for Mildred and Buddy to join Ollie in Germany. Ollie was among a group of fourteen officers who had been awaiting permission to reunite with their families. Mildred had applied for a passport in October, and Ollie requested permission for her to join him in December, but it took another three months for approval.[45] Mildred and Buddy crossed the Atlantic Ocean aboard a small army ship built in 1890 – the US Army Transport (USAT) *Buford*. In December 1919, the *Buford* gained renown for ferrying 249 leftists and anarchists being deported to the Soviet Union via Finland where they crossed a frozen landscape to be handed over to Soviet authorities.

Mildred later wrote of her journey:

> I, being a Kansas girl, knew nothing about ships except that they were necessary to carry me to my husband. I'd hesitate to take the Buford to cross the Mississippi, now. "Where ignorance is bliss. . ." Its capacity was forty passengers. Sixteen were Graves' Registration men, taking caskets to bring dead soldiers home from WW I. There were two dozen army wives, sixteen of us with babies, three months to three years old. It took us three weeks to get to Antwerp Boats that started after we did, got there before we did. An escort boat met the Buford to bring it into port. Ollie came out on it. Was that ever a joyous moment when I saw my grinning Ollie climbing up the ladder onto the ship.

Once Mildred and Buddy arrived, the Reeds set up house in the nearby town of Niedermendig, about twenty miles from Coblenz.

Niedermendig, Germany

In the beginning, relations between the occupiers and the local populace were poor. The doughboys who had fought in the trenches hated the "Huns," and the Germans resented the occupiers' hostilities, as well as the entire occupation. During the initial stages, doughboys with two or three overseas stripes (O/S) on their sleeves were rough on the local population and difficult to control; they especially bristled at taking orders from officers with only one stripe, or worse, no stripes. Germans, particularly decommissioned soldiers, resentful of the treaty terms and suffering from shortages of food and fuel, sometimes reacted to the American arrogance. Numerous fights occurred between disorderly elements of the U.S. Army and similar elements among the young demobilized German soldiers, most of which, however, originated in drunken brawls or jealousy over German girls.

The Reed family enjoyed good relations with the German people they met. Toddler Buddy learned the German language at the same time he was learning English. Mildred remembered: "When I'd tell [Buddy] nursery stories, he would go to the kitchen and tell them to the maids in German. When he talked German he would throw out his chest and gesticulate just like the Germans did. We had no doubt that he was the smartest child that God had ever made."

As it was impossible to import all of the essentials, Lieutenant Colonel Henry H. Sheen, chief of the AFG Quartermaster Corps, took command of a farm near Mulheim, about one hundred miles north of Coblenz, in April 1920 that had been established two months earlier as a vocational training center. The Quartermaster Corps put the farm to work to help meet the nutritional needs of the troops and their families. Being a Midwesterner who attended the Kansas State Agricultural College, Ollie was placed in charge of the farm and, on July 1, 1920, he was promoted to captain. The quartermaster farm functioned as both an agricultural school and working farm, supplying needed commodities for the army and families. The school provided Captain Reed with fifty-two students to work as farmhands. In a letter home, Mildred wrote:

> Since Ollie is a captain, he has been given more work. Also the use of a Cadillac and chauffeur. As I have told you, he has charge of the Farm, which supplies good milk (so many of the German cows are tubercular), chickens, etc., for American military personnel.

With the lack of fresh milk, Mildred would disguise powdered milk with cocoa. One day, Buddy saw some ducks dipping their beaks into a muddy pool. Looking at his mother, he exclaimed, "Ducks drink cocoa!"

Along with producing their own foodstuffs, the military relied on goods from local farmers available in the Quartermaster General Sales Store. This retail outlet for agricultural goods provided much-needed income to local farmers. They brought in fresh meats (except beef), poultry, fresh vegetables, eggs, milk, and flowers. Rabbits, pigeons, and bees were among the specialties of the Quartermaster store. Shortages of beef on the German markets were a result of disease and the in-kind reparation payments of livestock and goods to France and Belgium. Dairy products produced on the farm proved vital, as milk from local sources was scarce and of poor quality. Supplying fresh milk for the sick was a particular dilemma for the Army hospital, and the farm supplied enough for the infirmary and AFG families, with some left over for sale in the local market.

The extreme hardships and turmoil in Germany made travel in the country daunting, but the rest of Europe awaited American visitors. Major cities had been spared the devastation wrought upon countryside villages. Combat had reached the outskirts of Paris and Venice, but both cities, as with most European cities, had survived physically unscathed but economically in ruins.

Poverty was ever-present, an obvious sign of the war ravaged economies on city streets across the continent. Soldiers discharged from armies, and civilians whose countryside homes had been destroyed in the war, flowed into urban areas in search of opportunities. As post-war conditions improved in neighboring countries, Germany's economy continued its slide into depression and deprivation due to the grip of ongoing war reparations paid to the victors under the terms of the Versailles Treaty. Those stationed in the country needed a break and, for Americans, travel was cheap, so Ollie and Mildred took full advantage. Who knew when they might ever have the chance to see Europe again?

In a letter to friends at home, Mildred wrote about a European tour starting with a train ride to Paris and ending in Venetian moonlight:

> Our train left Coblenz at noon on October second, winding through the Moselle Valley for an overnight trip to Paris. The day we arrived was warm and sunny. Everyone was promenading. We were dismayed at the number of children we saw on the streets. What could the mothers be thinking of to let these little girls out alone in such a big city? We were disillusioned, however, upon a closer look at their faces, to find them far from youthful!
>
> So this is Paris! I suggested returning to the hotel and taking a tuck in my skirt. Really, the knees we saw were no better looking than mine. My husband restrained me. Most of the men reminded me of Charlie Chaplin and the rest made me think of villains, escaped from some play. I can't get used to the fancy beards the men in these countries wear. They look like they were someone else

in disguise.

In a French restaurant, we were introduced to hors d'oeuvres. We found them so fascinating and delicious, we made a meal of them alone. When we were served the entree, we said we had had sufficient and left. Crazy Americans!

Ollie and Mildred, right, in the Colosseum, Rome

From Paris, they traveled to Marseilles and Monte Carlo on their way to Rome, Florence, and Venice. In the same letter home, Mildred relayed her first encounter with gambling:

Monte Carlo is an interesting little resort on a spit of land reaching into the sea. Because Ollie was in uniform, he was not allowed in the Casino. I went in to buy some ivory and silver chips for souvenirs. I was awed - even frightened - by the tension in the air. Stony faced individuals were hovering intensely over the gambling tables. They threw withering glances at me when I asked the guard in a whisper where to buy chips. It was serious business for them. I thought gambling was supposed to be fun. How green can one be! I got out of there as quickly as I could.

Both Mildred and Ollie were rather unsettled upon their arrival in Venice. They were the wide-eyed kids from Kansas touring the Old World as it climbed out of war:

We were somewhat disillusioned at first when we got off of the train (which crept along on piles about two miles off of the mainland to have a dozen or so swarthy-faced scoundrels dash up and try to wrest our bags from us. We were certain they were robbers and cutthroats, but why were they allowed to attack strangers that way, right in the public station? We spurned them all and rushed forth to find the gondolier our geographies had described - with flowing sash, broad-brimmed hat and lyrical voice. To our dismay, we discovered the gondoliers were none other than our supposed villains. Dare we trust ourselves to their mercy? Since there was no other way to get anywhere we comforted ourselves with the thought of a revolver in Ollie's hip pocket and that we both could swim if worst came to worst. We stepped gingerly into the gondola.

We glided a mile and a half down the winding canal, but so busy were we looking right and left, it seemed but a short distance. Such queer houses, all built right on the water."

Finally, the couple relaxed into the beauty of Venice. They had left Buddy in the care of Leta, their nanny, back in Germany, and were enjoying this vacation as a second honeymoon:

It drizzled part of the time while we were there, but one evening was ideal and I shall never forget it. There was a full moon, beautifully reflected in the shimmering water. We engaged a gondola and glided into the night - now drifting in the shadows of the medieval palaces, now shooting silently out into the mellow moonlight. The world seemed as tranquil as dreamland. We had to pinch ourselves to make sure we were the same Kansas Jayhawkers we started out to be. Youthful laughter floated out to us from some balcony. As we turned into a narrow waterway, our gondolier's warning call, "Yoo Who," echoed and re-echoed musically on the still night air. Then hark! From down the stream comes the tinkling of guitars from a party of gay serenaders. Their accompanying voices raised in ballads and love songs made sweet melody. We recognized "Santa Lucia."

Feeding pigeons in St. Mark's Square, Venice

Romantic? I'll say it was. The gondolier didn't seem surprised when I snuggled into my husband's arms to be more readily accessible for his kisses.

Is it any wonder we were fascinated with Venice and left it with reluctance?

Good-night now. Love, from Mildred and Ollie.

✯✯✯✯

In June 1921, Ollie, Mildred and Buddy moved to Bendorf, where Captain Reed took charge of the city docks along the Rhine, primarily a cold storage plant that first housed some 900 tons of imported beef, and then became a gas and oil facility that handled 100,000 gallons daily.

The Rhineland region has traditionally relied upon natural resource exploitation and manufacturing as a foundation of the region's economy. After the war, barges of mining and metallurgical output returned to the Rhine en route to the Ruhr district, the industrial heart of Germany, stopping at the warehouses of Bendorf. The Bendorf docks and warehouses were central to the American presence, processing nearly two thousand tons of cargo every week.

In a 1921 letter home, Mildred described Ollie's responsibilities:

> Ollie's new job is in charge of a quartermaster depot here at the Bendorf docks. This is the storage and distribution place for food for Armed Forces in Germany. There's 120 thousand gallons of gasoline, a refrigeration plant, etc. Ollie is also town major for three of these little towns, which means arranging billets for soldiers to live with German families - and settling difficulties that arise therefrom. Ollie is the only officer here. He has a hundred or so soldiers. He also has 40 Germans in his employ. It is a big responsibility. The officer he replaced has been sent to the states to await trial for embezzlement. When Ollie was taking inventory, he found many empty cartons of cigarettes on the shelves. He remarked, "Looks to me like something crooked is going on here." A sergeant spoke up, "I had nothing to do with it, sir." That made Ollie suspicious. He began a thorough investigation, finding among other fraudulent things, a number of names on the payroll of people who didn't exist.
>
> The next night, we were awakened by flames at the warehouse and Ollie immediately thought of the gasoline exploding. He said, "If we can't get the gas tanks out onto the river, I'll send the chauffeur to take you and the baby and Leta [their maid], and whatever money we have around, as far away as you can get, because the town of Bendorf will be blown to smithereens."
>
> Was that a night to remember! Fortunately, they were able to get

the gasoline onto barges, and finally, brought the fire under control; but so much had been destroyed, further inventory was useless. Ollie realized afterward the refrigeration plant was as dangerous as the gasoline, if fire had reached it. Our guardian angels were on the job.

In Bendorf, the three Reeds moved into the home of a German officer. Along with the officer was his son, also a veteran of the war, a daughter, and the officer's wife. With army housing in short supply, German homes, as well as inns and other buildings, were required to open their doors to American troops. The men in the household were civil to the Reeds, but, as Mildred conveyed in a letter to friends, the wife and daughter did not disguise their disdain:

> They look the other way when we meet in the hall. I doubt if I would love my enemies, either, if they came in and took the best rooms in my house. On the second floor we have a parlor and a living room, hardwood floors, electric lights, fireplace, piano and nice furniture. It's a stone house with deep windows where Buddy sits by the hour playing with his tiny toy animals and watching activity in the street.
>
> On the third floor we have a spacious bedroom overlooking a big garden, beyond which, between the housetops and trees, we catch glimpses of the silvery Rhine and hazy hills that melt into the

"Unter den Linden"

skyline on the other side. Across the hall is our dining room (originally a bedroom), a neat kitchen, a room for our maid, Leta, and an improvised bathroom. The garden is the most attractive part of our new home. It is an acre of grass enclosed by vine-covered stonewall. Linden, fir, and fruit trees furnish shade. Flower bordered paths lead to a secluded nook where chairs and table invite you to linger.

There are roses everywhere, with violets and carnations lending fragrance to the air. Nearly hidden in the rushes, we find a pool with water-lilies (the Germans call them "sea roses"), and gold fish, which delight Buddy. In the center of the pool is a boy-and goose fountain. Sometimes on nice evenings Leta serves our dinner to us in the garden. Can you feature such luxury for two simple Kansas kids? We have to pinch ourselves to believe it.

Needless to say, many Germans resented being forced to house ninety percent of the American troops in their private residences for a daily pittance. And, at first, the occupying troops were not the most hospitable guests. Those who had fought in the trenches were haughty and antagonistic toward German civilians. War-weary soldiers and those coming from Prohibition Era America over-indulged in the freely accessible alcohol, causing problems both within homes and on the streets. As time went along, life settled into a routine and both sides grew increasingly appreciative of the other. Anti-fraternization orders were rescinded in September 1919 and relations between troops and the civilian population rapidly improved. Germans appreciated the American presence as a vital source of income – in dollars, and as a buffer against vengeful French troops and civilians, as well as German revolutionaries in the streets. When troop levels began drawing down, rail platforms began filling with crowds of German civilians bidding farewell. The annual report of the AFG for 1920-21 described one scene:

The sight of the throngs of Germans gathered about the train, of the sorrowful and in some cases tear-streaked countenances, an the shouted farewells made it difficult to realize that those leaving were soldiers of an army of occupation or that the crowds were composed of inhabitants of an occupied area. One could but reflect that the departing soldiers would probably meet with no such cordiality upon their arrival in their own country.[46]

In July 1921, as demands were building at home for withdrawal of U.S. forces from Germany, Major General Henry T. Allen, commander of American forces in the Rhineland, insisted that if a vote were held among soldiers and civilians, it would overwhelmingly favor the troops staying in place by ninety-nine to one.

Hunting party, Germany, circa 1921, Ollie in uniform, right of center, in uniform.

Conditions remained harsh for the German people. Reparations pushed and pulled the German economy in wild swings. Food supplies were low and prices high – and rising. Labor unrest was growing, with strikes targeting manufacturing and shipping.

While goods could be scarce, German unemployment was very low in the Rhineland; doughboys readily spent their pay in towns, helping to support local economies. As economic conditions continued to improve in the Rhineland, the French-inspired separatist Rheinische movement arose in 1923 with the establishment of a Rheinische

Republic, but it soon splintered, and a year later, it was gone.

Outside Rhineland, reparation demands were breaking Germany's economic back and the repercussions were felt throughout German society. Domestic politics in Germany were fracturing as the Communists allied themselves with the new Soviet government in Russia and ultra-nationalists rose in opposition. The new National Socialist Workers Party emerged in Bavaria, thanks in large part to its charismatic leader, Adolph Hitler, and his campaign against Communists, Jews and democratic principles, blaming all three for the woes of the post World War I country.

The Versailles Treaty brought an end to hostilities against Germany by the British, French and Belgians, but the U. S. Senate refused to ratify the treaty. A state of war remained in place between the two countries until July 21, 1921 when President Warren G. Harding signed the Knox-Porter Resolution ending hostilities with Germany, the Austro-Hungarian Empire, Turkey and Bulgaria. It took nearly another year for full diplomatic relations to be resumed when, on April 22, 1922, Alanson Bigelow Houghton, Ambassador Extraordinary and Plenipotentiary arrived in Berlin and was received by German President Friedrich Ebert.

With relations restored between the U. S. and Germany, domestic American support for the occupation evaporated, and calls for complete withdrawal gained ground. The German government, fearing interference by neighbors, particularly France, reached a compromise with the U.S. to maintain a force of twelve hundred men in the Rhineland until January 1922.

The reparations demands of thirty-one million gold notes every ten days was crippling the German economy. When they asked for an easing of payments, the Reparations Commission called the proposal inadequate, and added additional financial punishments with demands for greater authority over Germany's finances.

★★★★

By early 1922, the American drawdown picked up speed. Postwar domestic highs and lows drew America's attention home. The Roaring Twenties were prosperous and fun, while at the same time, the domestic economy fell into deep decline as Warren Harding took the office of president. There was little attention, let alone support, for American troops abroad. The League of Nations was the outgoing president's dream of a strong American presence on the world stage but American isolationism was rising. The U. S. Senate stood firm in its refusal to ratify the Versailles Treaty and its provisions for membership in the league, as the legislative body felt slighted that Wilson proceeded in treaty negotiations of such magnitude without consulting them.

When the occupation began in 1918, General Pershing marched 240,000 American troops into Coblenz. Midway through July 1919, American forces had been reduced to little more than 5,000 officers and 100,000 men. Over the next six months, these numbers were cut nearly eighty-five percent. Upwards of eight thousand doughboys and support staff were leaving Germany every day.[47]

On January 5, 1922, the USAT *Cambrai* ferried eleven hundred soldiers back home, leaving only fifty-five hundred American troops in Germany. The War Department ordered the number further reduced in February. President Warren G. Harding finally ordered the pullout to be completed by July 1, 1922.

By 1922, relations between the American troops and Rhinelanders had grown quite cordial. Non-fraternization rules were a thing of the past. Many American soldiers married German women during the occupation. The Americans were enjoying good food at cheap prices, a very favorable exchange rate, and the taste of a forbidden fruit back home – beer. At first, ready availability of alcohol for American soldiers was a problem but an AFG annual report later noted that "drunkenness on the part of Americans had become rare and, consequently, squabbles with the population were not the every-day affairs they had once been." The Americans had grown to embrace the entire Rhine region, and the region was embracing them in return. Unemployment for Germans in the Rhineland was very low, especially when compared with the rest of the country. The German government acknowledged the stabilizing influence of the Americans in helping to soften the social and political struggles of the day, and as a bulwark against the interference of outsiders.[48]

As the American withdrawal began, the French sought ways to increase their own hold on the region, something Rhinelanders feared most. French General Jean Degoutte proposed a plan to Commanding General Allen to station French troops in the American zone. Allen replied that the proposal was acceptable so long as the French respected standing regulations about "undesirable women and drink stronger than light wines and beers."[49]

On May 23, 1922, Mildred, seven months pregnant with their second child, along with Ollie, and three-year-old Buddy, sailed for the U.S. from Antwerp, Belgium aboard the USAT *Cambrai,* arriving at Ellis Island on June 2nd.[50] "We decided that it would be easier to travel with the baby 'in the oven,' so to speak, than with a tiny infant," Mildred later wrote.

On January 24, 1923, the last of the Americans marched out of Ehrenbreitstein Castle in Coblenz. Genuine sorrow could be seen on the faces of Germans and Americans alike.

Under the headline, "Der Abzug der Amerikaner" (The Departure of the Americans), The Coblenzer *Zeitung* newspaper published a lengthy editorial eulogy on that final day:

> Coblenz entered upon a new epoch of its history at the stroke of twelve today, when the star-spangled banner was hauled down from Ehrenbreitstein.

The newspaper looked back on the days when American troops entered the city, convinced that their intervention had turned the tide of war. The Americans carried animosities from the trenches but soon learned how much the two cultures had in common, the newspaper noted.

> It was not easy for the citizens of two nations until then wholly unknown to one another to grow accustomed to each other's habits while rubbing shoulders in the narrow confines of the Coblenz bridgehead.

The American soldiers, however, soon realised that the "Boche", who had been so much anathematised in their country amid the smoke and turmoil of war, was not the "Hun" that misguided public opinion in the United States had made him out to be.

After months, the American found that the Germans were not the deep-dyed villains as the world portrayed them during the war; that, as a matter of fact he was, at bottom, quite a likable, genial, industrious, peaceable and good-natured fellow.

The newspaper noted shared blood flowing through German and American veins in fostering mutual respect and understanding.

We thus hope that when the many thousands of Americans reach their mother-country they will make their better knowledge of the situation felt on matters political, and that America, so badly informed about Germany during the war, will be brought to a full realization of its interests and awake to the fact that Germany's liberty and welfare are as much a necessity to the United States as they are to Europe.[51]

As American forces departed, France assumed jurisdiction over the Rhineland. In response to Germany's default on reparation payments, the Reparations Commission imposed further sanctions on the industrial Ruhr District, allowing French troops to occupy and govern the industrial heart of Germany. Germans were unanimous in their opposition and resentment toward this heavy-handed takeover.

More than half of Germany's armed forces were war casualties, and although the country suffered little war damage, nearly half a million German civilians lost their lives to starvation brought on by wartime blockades and post-war deprivations, and the influenza pandemic. Those who had survived, veterans and civilians alike, became known as the "lost generation." Wounded, traumatized, and alienated by the horrors of war, a new level of discontent arose in post-war Germany. Tensions were high as political battles took to the streets. The revolution in Germany came to an uneasy end in 1919 with the establishment of the Weimar Republic that was quickly overwhelmed by hyperinflation. When Ollie arrived in the Rhineland, one dollar equaled about seventy marks. When they left, the exchange rose to ten thousand marks for a dollar, and hit one trillion marks a year later.

On October 30, 1922, less than three months before the American departure, Benito Mussolini, head of the nascent Italian Fascist Party, was sworn in as prime minister of Italy by King Victor Emmanuel. The National Socialist Workers Party of Germany, imitating the Italian black-shirted shock troops with their brown-shirted SA stormtroopers, readily engaging in bloody street fights.[52]

In November 1923, the Nazis failed in their coup, known as the Beer Hall Putsch, and Hitler landed in jail, but not for long. Der Führer and Il Duce were ascendant.

The seeds of the next world war had been planted.

Hitler addressing a rally of the National Socialist German Workers' Party in Munich, 1923. (USAHEC)

CHAPTER SIX

The Inter-War Years

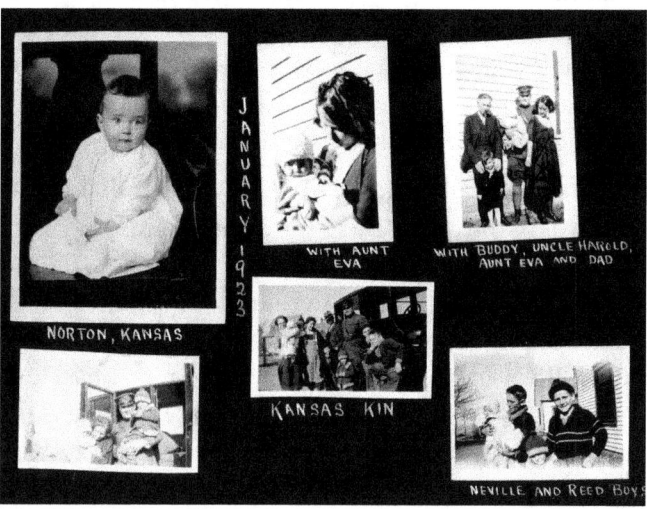

IN THE WAKE OF EVERY WAR voices clamor for reductions in forces and, most of all, in military budgets. The Great War was the War to End All Wars – so, worldwide conflicts were a thing of the past. What purpose would be served by maintaining a full-time military capable of fighting another world war? America had its priorities to tend to at home, not overseas.

Political compromises reached in the wake of World War I kept a standing army in place but its primary role would be defense of the homeland. There were lessons to be learned by the military from the last war to avoid repetition of the same mistakes. Development of new technologies and tactics required updated training.

Numbers in the ranks were in decline since the 1918 armistice, but Ollie remained in the army and on his promotion track. Many officers who served in combat suffered a demotion from their wartime rank, while those who served stateside retained their rank from the war years. Demotion encouraged many soldiers, enlisted and officers alike, to leave military service within a year after the armistice. Some left convinced that war had ceased for all time.

For Ollie, staying in the service meant that as soon as he arrived home from Germany, he and his family were on the move right away. Captain Reed received orders to report to Camp Meade, Maryland, for three months of training in the Tank School. Mildred, with her new baby nearly due, and Buddy, joined Ollie in Maryland where they rented a summer cottage along the Severn River in Sherwood Forest, today a popular gated community, less than twenty miles from Camp Meade, in an idyllic setting just outside of Annapolis.

"The quarters at Camp Meade, where Ollie was ordered, were impossible," Mildred wrote. "There were holes in the floors and walls, no plumbing, no screens. The weather was beastly hot. The baby was due in a few weeks. Ollie took the five hundred dollars we had earmarked for payment on a car, and rented a cottage in a summer resort on the Severn River in Sherwood Forest near Annapolis. He came home on the Interurban several times a week.[53] It was ideal for us. We give credit for having such a good baby to the peaceful surroundings."

Infant Ted Reed

On July 25, 1922, Theodore Harold Reed was born at Walter Reed Army Hospital. Mildred spent an agonizing eighteen days in the hospital with Teddy's birth. The worst part was being away from three-year-old Buddy, but Leta, the family's nursemaid from Germany, was with them to help.

From 1916 into 1919, rising from private to lieutenant, Ollie had learned the basics of the wartime military and its training regime. As a lieutenant and captain, he participated in one of America's first peacekeeping missions. Now, in 1922, he was returning to the United States to begin training in the new army, starting with the Tank School.

Tank warfare, adopted by the British in World War I, was replacing the horse cavalry. Commandant of the Infantry Tank School Colonel Samuel D. Rockenbach promoted integrated training:

> The modern Infantry is technical Infantry, and that its personnel must know the automatic rifle, the machine gun, the accompanying cannon and the tank as thoroughly as the rifle was known.

> In the World War, destructive power had increased a thousand fold over that of the time of Napoleon and of our Civil War... Means to overcome machine guns, wire entanglements and trenches without exhaustion had to be devised. Hence, the tank was devised to move independent of roads, crush obstacles and carry protected gun power into and beyond the enemy's lines.
>
> In the next war, machines are going to supplement the efforts of flesh and muscle to a larger extent than ever before...[54]

Tanks were introduced into combat by the French in 1917 with their Renault FT. Captain George S. Patton employed armored Dodge vehicles in the Mexican Expedition and was the first officer assigned to the American Tank Corps under the wartime command of Brigadier General Samuel D. Rockenbach with the AEF in France. Following the war, Patton was assigned to the Tank School at Camp Meade, where he met fellow captain Dwight D. Eisenhower. The two men believed deeply in the critical role of tanks in combat, going so far as to write a pair of essays promoting mechanized warfare, running contrary to the higher ranks and political thinking that had incorporate the Tank Corps into the Infantry as part of the National Defense Act of 1920. The Cavalry would argue for years to come about command and control of the mechanized units but in 1920, theirs was a lost cause.

Captain George S. Patton (National Archives)

Writing in the May 1920 issue of *Infantry Journal* using his wartime rank, Colonel George S. Patton, Jr., Tank Corps and simultaneously in *Cavalry Journal* with his then-current rank, Major George S. Patton, Jr. Cavalry, Patton led his essay, "Tanks in Future Wars," with Pattonesque dryness and sarcasm:

> In view of the prevalent opinion in America that soldiers are, of all persons, the least capable of discussing military matters and their years of special training are as nothing compared to the innate military knowledge of lawyers, doctors, and preachers, I am

probably guilty of a great heresy in daring to discuss tanks from the viewpoint of a tank officer.

I am emboldened to make the attempt, however, not from a bigoted belief in the infallibility of my opinions, but rather in the hope that others will assail my views and that the discussion thus engendered will in a measure remove the tanks from the position of innocuous desuetude to which they appear to have been relegated by the general public.[55]

Patton was writing at precisely the time the National Defense Act of 1920 was being enacted, reducing the size of the military and subordinating the tank corps to the infantry.

In the months following enactment of the defense act, Captain Dwight D. Eisenhower wrote a more reasoned essay, with a pithy title, in the *Infantry Journal*, "A Tank Discussion."[56] Both young officers made compelling arguments for the future of mechanized warfare, and both were called on the brass carpet for their public disagreement with prevailing policies.

Patton and Eisenhower had moved on from the Tank School by the time Ollie arrived, but Camp Meade and the Tank Corps were under the command of now-Colonel Rockenbach, the wartime commander of the Tank Corps in France. The Tank School was staffed with twenty-seven officers and 134 soldiers operating twenty-five light and heavy tanks in the training program. Officers and enlisted men were trained in the operation and maintenance of the armored machines, and in crew training. The program employed both the heavy Mark VIII tank, modeled after the British Mark V, and the M1917, fashioned directly from the French Renault FT17. The training program for a full-time tanker would last from six to nine months but for officers, like Captain Reed, who were receiving just an introduction, the course lasted three months.[57]

Besides dismantling the Tank Corps and assigning the armored brigade to a subordinate role under the Infantry, the National Defense Act of 1920 also updated the 1916 act by setting the manpower levels for the standing military in the post-war period and establishing three components of the army: the Regular Army, the National Guard and the Organized Reserves. The Regular Army was assigned a maximum strength of 17,726 officers, more than three times the pre-war levels, and an enlisted strength of 280,000, with actual levels dependent upon Congressional appropriations.

In the first month of demobilization following the war, the Regular Army was reduced by 650,000 officers and men. Within nine months, over three million men were mustered out of service. By the end of 1919, the army had been reduced to 19,000 officers and 205,000 enlisted men – an all-volunteer Army.

In January 1921, Congress further shrunk the enlisted ranks to 175,000 by reducing appropriations, and six months later took that number down to 150,000. The next year, Congress limited the Army to 12,000 commissioned officers and 125,000 enlisted men, a number that would hold steady until 1936.[58]

★★★★

With Ollie's ninety-day tank training completed, the Reed family was again on the move, this time making only a short journey north from the Severn River to Fort Howard, known as the Bulldog at Baltimore's Gate, on the Chesapeake Bay, for three months of coastal artillery training during the autumn and winter of 1922-23.

Mildred thoroughly enjoyed living on the Chesapeake Bay, except for one incident: "What a thrill it was to see sailboats from our bedroom window! How we enjoyed fresh seafood! After feasting on crabs for a week, we found a man's half-eaten body under the dock. That diminished our enjoyment of crabmeat for awhile."

Ollie's tour at Fort Howard finished in January 1923, so the Reed family embarked on a wintertime trip to Kansas. This marked the first time they had visited Kansas since Buddy and Ted were born, and the first they had been home as a couple since they were married. Mildred remembered, Ollie was especially pleased to show Dr. Kenny [Mildred's doctor when she had tuberculosis] how well he had taken care of her.

> I was ten pounds heavier, and an inch and a half taller than when I was married. (I had weighed 92 and Ollie 184, then.) Dr. Kenny had told folks I wouldn't live more than two years with that soldier boy dragging me around the country. I'm sure that no one in Norton County had had as easy and pleasant six years as I!

Buddy, left, and Teddy Reed, 1923

Mildred had adopted some "exotic" ways from her foreign travels, particularly in the manner in which she dressed young Theodore: "Teddy was a chubby, happy baby, with the prettiest, dark, curly hair you ever saw. I had lovely embroidered dresses from Germany, and lace bonnets from Italy. I was not about to waste them. Folks would say, 'What a beautiful little girl!' I didn't say a word. When he was two, his father said, 'All right now. We cut this boy's hair and put pants on him.'"

Ollie received his next orders, a step up the command ladder, to the Infantry School Company Officers' Course at Fort Benning, Georgia.

Fort Benning is situated on the banks of the Chattahoochee River just south of Columbus. Both the fort and Infantry School were practically brand new in 1923, having opened only five years earlier in 1918, with full classes beginning the next year. The Infantry School originated as the School of Musketry at the Presidio of Monterey, California, in 1907, relocating to Fort Sill, Oklahoma, in 1912 before finding a permanent home at Fort Benning.

Love and Sacrifice

In the 1924 Fort Benning *Doughboy*, the anonymous "uncensored diary of Nosmo King, sometime Captain of Infantry" recounted his first look at Fort Benning:

> Took the bus to Fort Benning, and saw much dry ground, two creeks, cotton fields, tumbledown shacks, and pine woods; and the fort looming up on the hills above the river. The first impression is that of a reformed cantonment of the vintage of '17.[59]

Fort Benning stood as an incomplete work in progress. Even during the war, Camp Benning units were transferred to other facilities and new construction halted in an effort to downsize the installation. With the war's end, members of Congress, spurred by the cost savings in military spending and the opposition by local residents to expanding the camp, tried to shutter it completely. The army had declared Benning to be the home of the new Infantry School in the waning days of the war, but Congress was not convinced. The Senate Subcommittee on Military Affairs held hearings on the reorganization of the army in 1919, during which Benning came perilously close to being closed. Going into the hearings, Senator Hoke Smith of Georgia was a proponent of reducing the size and importance of Camp Benning, but by the end, he supported the existence and expansion of the encampment, carrying the day.[60] [61]

Flooded entrance to Ft. Benning (Donovan Research Library)

On February 20, 1920, Congress passed a lifeline to Camp Benning and the Infantry School, but funding was not approved until the following year.[62] Without a budget, essential projects such as completing the water and sewer works advanced with whatever materials could be found.

In 1921, a narrow-gauge railway, known as the Dinky Line, the Chattahoochee Choo-Choo, or Old Fuss and Feathers was moved to Camp Benning from the Davenport, Iowa Locomotive Works where it had been used during the war. Its twenty-seven miles of track around the base proved vital to the construction of new facilities, as well as carrying men and material with ease.⁶³

In 1922, Camp Benning became Fort Benning, and improvements continued by the time the Reeds arrived in 1923. New officers' quarters had been built and the first phase of a new hospital complex was opened in October 1923. The introduction to the 1923 Infantry School yearbook, *The Doughboy*, summarized the development of the school:

> When in November 1881, the General-in-Chief regarded "as admirable" the "new School of Application at Leavenworth," the Army unconsciously entered upon a period of Rennaissance [sic]. Later General Wagner, like Erasmus with the classics, culled the best military thought of Europe and America and presented it to the line officer.⁶⁴ The profession of arms rudely but surely started upon a larger development of science and skill. The Army stretched itself to find that it was awakening from the Dark Ages of provincial life into which the nation had thrown it. The Infantry School is the culmination of the Rennaissance [sic] of the United States Army.⁶⁵

The new commandant, Brigadier General Briant H. Wells, spearheaded the design of a new concept in military installations, emphasizing the outdoor environment of the entire post, adding sports and recreation facilities that were welcoming to soldiers, officers, and their families.

The inspiration for Wells's designs was the City Beautiful Movement championed by landscape architect and city designer, Frederick Law Olmsted, and his designs at the 1893 Chicago World's Fair, known as the World's Columbian Exposition. The movement developed some of the most distinctive urban designs at the end of the nineteenth century and beginning of the twentieth, many of which remain in place today, such as Central Park in New York, Michigan Avenue in Chicago, as well as the monumental layout of Washington, DC and its Beaux-Arts style architecture. After the war, the army needed to stem the loss of trained personnel and decided to spur the development of attractive environments, highlighted by recreational facilities, as an effective method for retaining experienced soldiers.

The first undertaking was major landscaping and a central wheel-and-spoke design unifying the campus with the Main

Post Cantonment as the hub. New housing, bridges, hospital and recreational facilities were all built, with materials hauled by the Dinky Line into place. Area workers were hired, helping the local economy, while troops supplemented the workforce.[66]

Mistakes were made. Officers' quarters were built with sharply gabled rooftops, better suited to heavy snowfalls and certainly not for Georgia summers when the upper floors became super-heated. The army's Construction Division had mistakenly sent the plans for housing at Fort Ethan Allen in Vermont to Fort Benning. Regardless, the new officers' quarters were regarded as palatial by the former shack-dwellers of Camp Benning.

Although General Wells's building and beautification scheme was never put into full effect as Congress tightened the inter-war budget strings, it became the basis for future planning; many of its features are part of the post today.

Fort Benning was destined to become the largest military school in the world, at a time when the training of infantry officers was undergoing tremendous changes.

In the National Defense Acts of 1916 and 1920, the Army continued its slow move away from the ragtag collection of militias to an organized standing army, in spite of political resistance, and the hard lessons of hurriedly developing the officer corps for the war that had just ended. The Infantry School would be the backbone of continuing army officer training.

The school was founded upon the principals of training by demonstration as laid out by Baron Friedrich Wilhelm von Steuben 150 years prior, utilizing a body of troops "for the purpose of forming a corps, to be instructed in the maneuvers necessary to be introduced into the army and to serve as a model for the execution of them."[67]

General Pershing, writing in the introduction to John McAuley Palmer's *Washington, Lincoln, Wilson – Three War Statesmen*, pointed to von Steuben's influence on Washington:

> If our fathers had followed the scientific plan so carefully elaborated by Washington with the aid of Baron von Steuben and his other generals, we should have been better prepared in the

beginning for the War of 1812, the Civil War, and the (first) World War.⁶⁸

The introduction to the 1924 yearbook, The *Doughboy*, framed the resolve of the Infantry School:

> The Infantry School is dedicated to the Infantry and exists by and for the Infantry. The spirit underlying the institution is the same as that back of the Infantrymen, which spirit, to quote the words of an eminent and gallant soldier, himself a Doughboy."
>
> In response to America's call, wrote into the history of the World War an immortal record on the battlefields of France, winning at a cost of 89 percent of all American dead the greatest victory which has ever crowned the achievements of American arms.
>
> Which will continue by its willing and fearless acceptance of hardship and sacrifice to preserve all that is manly and noble in the military profession, and to insure to America the integrity of her splendid institutions whatever the source from which they may be threatened.⁶⁹

Course work in the school was divided almost equally between classroom and field. As technology advanced weapons systems, greater emphasis was placed upon weapons instruction. In June 1923, ten-day field maneuvers were introduced into the curriculum and immediately became a centerpiece of the program.

The Infantry School hosted four courses of study for various ranks in service. Officers attending the school ranked as high as the six brigadier generals during the time Ollie was enrolled. There was a Refresher Course to help ten colonels over ten weeks "clear the cobwebs" and prepare for taking command of a regiment at the completion of the course. The Advanced Course had seventy-three colonels, lieutenant colonels and majors, running from September 1923 to May 1924. Captain Reed was one of 163 captains enrolled in the Company Officers' Course from October 1923 to May 1924, learning leadership at the platoon and company levels.

Daily life for the officer-students during the nine months of the Company Officers' Course was a steady diet of lectures and monographs about history, tactics, and psychology; turns at the range

Love and Sacrifice

for pistol and rifle marksmanship - with an ever-present bayonet at one's side. It was a grueling course; some called it torture, so the men had to find lighthearted moments when they could.

Calculator. (National Infantry Museum)

One of those came in the form of a dog, an old mongrel who had wandered onto the base at the very start of the program – Calculator. He was described as a mutt, but of the highest order of intelligence.

Cal or Calc, as he was called, would ride the train into Columbus or, if he missed the train, hop into an officer's car to hitch a ride to wander the city streets in search of whatever handouts he could find. At the end of the day, an officer would pick him up at a regular location for the ride home. Calculator's home was wherever he curled up at the end of the night, regardless of the host's rank.

As an item in the *Infantry School News* noted for incoming students: "Cal has no home for every place is his home, he is everybody's friend and everybody is his friend. You officers coming to school next year will soon know Calculator."

Calculator was so named by feigning a limp when begging for food: "He puts down three and carries one." There was a standing column in the base newspaper featuring the Great Dane seeking wisdom from Calculator, and of their relationship with the Grey Squirrel and Blue Bird

On August 23, 1923, Calculator died, a victim of strychnine poisoning by a perpetrator who was never apprehended. All of Fort Benning and those who had served there mourned. Newspapers around the country carried the story of Calculator's death and contributions poured in for a Calculator memorial.

The front page of the September 7, 1923 *Infantry School News* ran a eulogy to Calculator with the installation of a

memorial plaque:

> Pick up any *Doughboy* since the start of the Infantry School, and you will find Calculator featured on its pages. This faithful little pal has been as much a part of this place as the trees. In years to come, he will be a tradition, remembered by some and forgotten by others.
>
> Let this monument stand as a permanent testimonial to the admonition that unkindness to dogs will not be tolerated by the Infantry. Our late President Harding took an oath never to kick a dog; and this monument will stand as a pointing finger of warning to those inhuman morons who throw all the dictates of love and affection to the winds and kill one of the best friends man has, the dog - a real pal.
>
> Let this monument stand, too, as a testimonial to the love and affection with which the Infantry School regarded dear old Calc -- let it be symbolic of a permanent half-mast in honor of a dead companion.

Today, the memorial to Calculator stands, embedded in a large stone, at the National Infantry Museum in Fort Benning. The engraved homage, attributed to Colonel Paul Giddings of the 13th Infantry, simply reads:

"He made better dogs of us all."

Calculator memorial at Fort Benning.
(Donovan Research Library)

Ollie's favorite diversion at Fort Benning was football. The *Infantry School News* of April 11, 1924, talked about the upcoming spring football season at Fort Benning with Ollie assisting, as well as playing for, head coach "Shrimp" Milburn, otherwise known as Major Frank W. Milburn, who later served as lieutenant general in command of the V Corps in World War II and Korea:

> Ollie Reed will be remembered as the giant tackle of last year's Infantry Varsity. He was field captain of the eleven and led them through a strenuous and at times disastrous season. Under his

guiding hand the spring candidates will get all the work they want in the short session before summer sets in.

Fort Benning's sports and recreation programs were outstanding. Year-round sports included football, baseball, basketball, lacrosse, boxing, track, polo, tennis, golf, hunting, and swimming. All of the sports programs were self-sustaining without any financial support from the army, but participation in football and polo were seen as proving grounds for an army career.

The Infantry School had varsity teams in all sports and was an honorary member of the Southern Intercollegiate Athletic Association. Their competition included Georgia, Auburn, Georgia Tech, Oglethorp, Florida, Loyola, Mercer, Gordon, Vanderbilt and Sewanee. The people of Columbus flocked to home games in an old wooden stadium. Local merchants enjoyed handsome profits from the crowds coming into town. In 1924, the "President's Cup" was introduced, and the Infantry School represented the Army for the first three seasons; in 1926, with Major Dwight D. Eisenhower as one of four coaches, an All-Army team was assembled and trained at Fort Benning.[70]

Life was very quiet for Mildred and the children during their six months living in Columbus, Georgia. Buddy turned five and Ted had not yet turned two by the time they left. The boys kept Mildred busy while Ollie was in school. Although they were only renting the house at 2714 Beacon Street, a small stucco bungalow, Mildred was fond of it.

> I had dreamed of having a fireplace - nobody in Norton had them when I was young. Pictures of them looked so romantic. The house we rented in
Columbus had a fireplace in every room. If we faced the fire, our faces blistered and our backs froze. If we stood in front of it, our backs roasted while our noses froze. I settled the situation by moving the overstuffed davenport in front of the fire, with chairs boxing in the ends and blankets to keep out drafts. I climbed into my nest with my little boys and there we played all day long.

Mildred and Ollie at Fort Benning.

It was in Columbus where Ollie was determined to teach Mildred how to drive.

Ollie decided that I should learn to drive the Nash he had purchased shortly before we left Maryland. I was contented with my home and babies, and had no desire to be dashing around the countryside in an automobile. He took me out for a lesson. I didn't catch on as quickly as he thought I should. A woman, driving along like the Queen of Sheba in a pickup truck, passed us. As we ate her dust, Ollie said, "Now, if a woman of her intelligence can drive a car, surely you can."

I said, "Do you see that stone wall beside the road? I'm driving your damned old car right into it." He stopped the car six inches from the wall. For some reason, he didn't say anymore about me learning to drive.

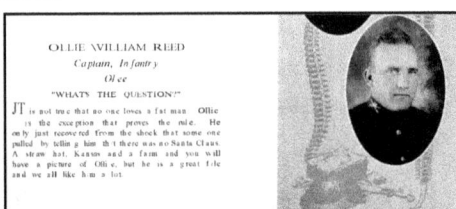

Ollie graduated from the Infantry School in 1924, and that year's *Doughboy* poked good-hearted fun at the man called "Olee":

> It is not true that no one loves a fat man. Ollie is the exception that proves the rule. He only just recovered from the shock that some one pulled by telling him that there was no Santa Claus. A straw hat, Kansas and a farm and you will have a picture of Ollie, but he is a great file and we all like him a lot.

After graduation, most of the 231 Regular Army graduates of the Infantry School returned to their old units. Ollie received a new assignment, one that would use his demonstrated strengths of teaching and leadership – professor of Military Science and Tactics, and commandant of the Reserve Officer Training Corps at Drexel Institute of Technology in Philadelphia.

CHAPTER SEVEN

Professor Ollie Reed

Military Science and Tactics

OLLIE W. REED, Capt., U. S. A.
Professor of Military Science and Tactics

HENRY E. KELLEY, First Lieut., U. S. A., *Asst. Professor of M.S. and T.*
GUSTIN McA. NELSON, First Lieut., U. S. A., D.O.L., *Asst. Prof. of M.S. and T.*
ADAM F. TEPPER, Sgt., U. S. A., *Asst. to Prof. of M.S. and T.*

We bought a stucco house in an old orchard in Merchantville, New Jersey, across the Delaware River from Philadelphia. We were only a half an hour from the college that was closer than any of the Pennsylvania suburbs we investigated. Teddy called it the "tucco house". We bought it very reasonably, because it wasn't finished. We dug the sewer ditch, shingled the roof, papered the walls, painted the woodwork, and planted our thirteen pear trees to make a garden. All this labor made it really ours.

REMAINING ON ACTIVE DUTY, Captain Reed was next assigned to the Drexel Institute of Technology in Philadelphia as professor of Military Science and Tactics.[71]

The Drexel Institute, today Drexel University was founded in 1891 by Francis Martin Drexel and his son, Anthony Joseph Drexel, both wealthy Philadelphia financiers. The school was established to train "young men and women in skills to meet social, commercial, and industrial change; the while providing for them and for the people of Philadelphia new opportunities for cultural development by means of library, museum, art, music, lectures, and evening classes."[72]

Dr. Hollis Godfrey, a native of Lynn, Massachusetts, educated at Tufts College, Harvard University, and the Massachusetts Institute of Technology, was named president of Drexel in December 1913 and served until his death in 1936. When war approached and enrollment dropped as young men joined the mobilization effort for the First World War, Godfrey sought to foster collaboration between Drexel and the military.

Godfrey viewed the nation's growing interest in military service as an opportunity to bring education and industry together, improving technical education to the benefit of the country and its national defense. His 1917 keynote address at the twenty-fifth-anniversary convocation was titled, "The Service of the College to the State," gathering speakers to reinforce that message. Among them, Dr. Samuel P. Capen, chancellor of the University of Buffalo, emphasized the "vast and untapped resource of potential military usefulness had been opened up to the government by the mobilization of research workers, and of the university, college and school laboratories." Capen urged that engineering schools follow the plan, which had already been initiated by some colleges, of organizing courses in military map-reading and surveying, bridge building, telegraphy, radio, and machine servicing.

"This one cry, the mobilization of the technical power of the nation as a war measure," the *Philadelphia North American* declared, "was carried through the whole program and emphasized by each speaker."

In 1919, two years of military training became compulsory for all male students in the School of Engineering. The Reserve Officers' Training Corps at Drexel was approved by the War Department and established on January 7, 1919 under Lieutenant James P. Lyons. Drexel prided itself in serving America on the assembly line and on the firing line.[73]

Chancellor Godfrey's interest in military matters predated his tenure at Drexel. He was perhaps best known for his 1908 book, *The Man Who Ended War*, a science-fiction story where the U.S. Secretary of War, along with the war ministers of other nations, received a letter ordering him to destroy his weapons and disband his military, or suffer dire consequences.

"It is a declaration of war against the civilized world in the interests of peace," the secretary said after reading the letter.[74]

In their January 19, 1936 obituary of Dr. Godfrey, the *New York Times* wrote of his literary work: "His ingenuity in conceiving the mechanical devices of future wars displayed in the book, six years before the beginning of the World War, led to his working with Elihu Root, General Leonard Wood, and Howard Coffin in creating the Council of National Defense of the United States. From 1916 to 1918, he served as federal commissioner for the advisory committee of the council and was in charge of its section of engineering and education."[75]

THE R.O.T.C.
1926 Drexel ROTC (Drexel University Library Archives)

When Captain Reed took charge of the ROTC at Drexel in 1924, it was rated as average among other ROTC programs of its day. In his first year, only five ROTC students received commissions. By the end of his fourth year, twenty-eight men were commissioned into the Reserve Officer Corps. In his first two years, the program failed to exceed its quota of thirty-five applicants, but by the time he left there were an average of 140 applicants each year. By the end of his four-year term, from 1924 to 1928, Reed ensured that Drexel's ROTC program steadily maintained an excellent rating. Among Reed's accomplishments at

Drexel was organizing the Reserve Officers' Training Corps Band in 1925 that, by 1942, had grown to more than sixty members. Today, the Drexel Army ROTC program, known as Task Force Dragon Philadelphia, graduates roughly the same numbers of commissions as it did when Captain Reed was in charge.

Ollie treated the students with respect, providing a guiding hand that firmly held them to their course. In return, students had a great deal of respect for the army officer who demonstrated a quiet knowledge of military history and tactics. Mildred was popular among the students as a mother and friend to all, just as Captain Reed was their father figure. He brought the lessons learned at Fort Benning and put them to work, emphasizing experiential learning. The college rifle clubs grew as students learned marksmanship. There were regular visits to Army installations and maneuvers for the students. Captain Reed continued the philosophy of learning by doing, not by the book – and his students responded.

1926 Drexel football team; Ollie 3rd row from bottom, 4th from the right.

Ollie's love of football came in handy when, in 1926, Drexel suspended full-time professional coaching at the school and returned to faculty coaches. Captain O. W. Reed stepped up to coach the varsity football squad, assisted by Lieutenant H. E. Kelley and Dr. E. J. Hall. Training opened with two players suffering broken collarbones, but the season started on a more favorable footing as Drexel defeated Georgetown University 42-0. Injuries continued to plague the team as they dropped their home opener to Susquehanna 21-0 two weeks later. The 1927 yearbook, *The Lexerd*, praised Coach Reed and his staff:

> The season was not as good as was expected with a veteran team, but many injuries hurt the team's chances. A great deal of credit is due to Coach, Captain Ollie Reed and his able assistants, Lieutenant Kelley and Dr. Hall for the spare time and services they gave the team.

In 1927, the school returned to professional coaches, recruiting Walter H. Halas, the athletic director at Notre Dame University and the brother of American football pioneer George Halas, to coach not only football, but also basketball and baseball.

Captain Reed and the Drexel Girls' Rifle Team.   Drexel ROTC Band. Photos: Drexel University Library.

Besides launching the very popular ROTC Band in 1925, the following year Ollie directed the first Military Ball at Drexel. The ROTC program had an officer corps standing at thirty-two members, ranking from colonel through second lieutenant. The combined men and women's rifle teams grew to two hundred members from twenty-six in 1923. Captain Reed introduced his cadets to summer maneuvers at off-campus locations, such as Camp Meade. He was regarded as a father figure, stern and tough, but fair and a beloved mentor to the young men and women, many away from home for the first time. Mildred opened her home across the Delaware River to students who were equally thrilled when she brought her sons to the campus. Over four years at Drexel, Ollie revived a moribund ROTC program at the college. As it was written in the 1926 Lexerd yearbook:

"With all due respect to the units of previous years, we do not feel boastful in stating that this year's unit is the largest, best and snappiest one that Drexel has ever turned out."

Mildred loved her Craftsman-styled "tucco" house at 103 Orchard Avenue in Merchantville, New Jersey, and enjoyed watching her boys grow and enter school.[76] It was a brand new house when they purchased it in 1924 and the Reed family was growing into the neighborhood along with their neighbors. But, Ollie remained on active duty and the Army called upon him to move again

The May 28, 1928, front page of the *Drexel Triangle* student newspaper brought news of Captain Reed's forthcoming departure the next month:

No one can or will ever forget Captain Reed. His service to the school and his never-ending interest in the students and student problems of Drexel assure tender recollection of a man who thought never of himself but always of others. Some may remember, perhaps, how as Freshmen they were herded into a large room where the Captain reviewed the "rookies" and instilled instantaneous admiration and respect.

Those with whom Captain Reed has come in contact will ever remember him as a staunch friend, a just disciplinarian, and a good fellow to boot. Then, too, the entire student body has had the sincerity and warm-hearted interest of Mrs. Reed in her own home, or assisting in the social life of the college and who has shared Captain Reed's human interest in the students.[77]

With his Drexel tenure at an end and owed some leave time by the army, Ollie's native feelings of wanderlust returned as he considered following a dream of joining friends on a trip around the world on a freighter. "He was feeling sorry for himself that he was tied down with a family, and would never get to do it." Mildred later wrote:

> I said, I admit I caught you pretty young, but I'll tell you what I'm going to do. I'll give you a year off. You have three months leave due you. You can get it extended as educational leave. The boys are small enough that I can manage them. We're settled in this house and community. I'll get along all right. Look into it. You're only 32. Young enough to still have adventures. Maybe you could write them up and make a million.

Ollie met with the freighter captain and inspected the boat, returning home furious that night, telling Mildred, as she recalled:

> I am so mad at you!! When I was young, I wouldn't have minded that greasy galley, and the awful place to sleep and the stinkin' toilet. But you have gotten me accustomed to nice things and clean places. I would have loved the whole freighter when I was a kid, but now it was repulsive. It's all your fault.

Instead of a round-the-world adventure on a tramp steamer for Ollie, the Reed family headed back to Fort Benning and Columbus, Georgia.

For the first time, Mildred had regrets about leaving a place:

> I could hardly bear to leave our little home in Merchantville. I knew we'd never own another house... I especially hated to leave the casing to the basement door where we had marked the quarterly growth of the boys. As we drove away, I was sobbing.
>
> Good old Pop said, "If you will quit crying, I'll stop at Strawbridge and Clothier on the way through Philadelphia and you can buy the Haviland China like your mothers, that you've been wishing you had." That dried my tears immediately.

CHAPTER EIGHT

Back to Benning

The Reed Family, December 1930. From left: Ted, Buddy, Mildred, and Ollie.

AFTER FOUR YEARS of leading students and young cadets at Drexel, Captain Reed returned to active duty with the 29th Infantry Regiment at Fort Benning. Under the motto "We Lead the Way," the 29th was designated a Demonstration Regiment for the Infantry School at Fort Benning.

With the introduction of field maneuvers into the Infantry School curriculum, experiential learning by doing continued to be a critical component of the officer-training course. Accompanying classroom learning and weapons practice on a controlled range, officers were tasked with tactical planning for an assault or defense, and soldiers were in the field to bring that planning to a successful, or unsuccessful, conclusion. Weapons' training was another integral part of the Infantry School program. Officers enrolled in the school had the opportunity to learn about the full range of weapons systems, their operation, and management. The weapons companies played a significant role in both aspects of Infantry School education – in actual field use during maneuvers, and by example in daily training exercises.

Since Ollie left Fort Benning in 1924 for Drexel, the fort, and Infantry School had been growing rapidly, converting ninety-seven thousand sprawling acres of red mud and sand into a premier training academy for infantry troops and officers. Up to 1930, the Infantry School had graduated 5,064 officers into the ranks of the Regular Army, National Guard, and Reserves.

Brigadier General Edgar T. Collins succeeded General Briant Wells as commandant during Ollie's first stint at Fort Benning. In 1926, he appointed Lieutenant Colonel George C. Marshall as assistant commandant and head of the Infantry School the following year. Each Infantry School commandant brought their own unique philosophy and agenda to the post. Marshall, who would go on to become Army Chief of Staff during World War II and carry out the Marshall Plan in its wake, wanted to invigorate the education and training program, bringing changes that became known as the "Benning Revolution."[78]

Lt. Col. George Marshall
(Donovan Library, Ft. Benning)

Marshall recruited innovative young officers who had demonstrated their capabilities in World War I. Marshall's team overhauled the curriculum at the Infantry School, emphasizing high-quality instruction and innovative execution. The instructors brought to Fort Benning by Marshall reads like a Who's Who of Army leadership in the forthcoming world war: Lieutenant Colonel Joseph W. "Vinegar Joe" Stillwell as Liaison Officer and chief of the First Section; Captain J. Lawton "Lightning Joe" Collins, an instructor in the First Section; Lieutenant Colonel Morrison C. Stayer, chief of the Second Section who would go on to win two Distinguished Service Crosses in World War II as a Major General in the Army Medical Corps; Captain Stonewall Jackson was an instructor in the Second Section;[79] Major Omar N. Bradley, chief of the Third Section, would rise to five-star general leading American forces in Europe; and Major Edwin F. Harding, chief of the Fourth Section, later led the 32nd Division in the Pacific as a major general through 654 consecutive days of combat in World War II.

Tactical improvisation and creativity were hallmarks of Marshall's program, emphasizing training geared specifically to prepare an officer in combat to coolly assess any situation, no matter how chaotic, and to make clearheaded decisions. Marshall visited field exercises to explain his view of the day's obstacles. He then might select a student to act in command as the others provided their critique. He wanted students to think confidently on their feet.[80]

"He would come in and take all your maps, take all your notes, and see how well you did without them. The ones who could be put under that kind of pressure and scrutiny and then succeed, Marshall felt like they had achieved the goal. He kept a little black book and would jot down whom he felt the leaders of tomorrow's Army would be," Zachary Frank Hanner, National Infantry Museum director, told the authors of *Fort Benning: The Land and the People*.[81]

By 1928, the construction that was just beginning when Ollie was first assigned to Fort Benning in the early 1920s, using designs by innovative urban planner George B. Ford, was in its final phases. Among the buildings completed in the 1930s were the Post Chapel, Officers' Club, and the original Infantry School Building, now headquarters for the Western Hemisphere Institute for Security Cooperation (WHINSEC), formerly the School of the Americas.

The turnover in commandants over the decade had disrupted smooth planning, but that changed in May 1929 when Brigadier General Campbell King received the support of the War Department to stabilize planning and harmonize features, resulting in a graceful symmetry in the overall design of Fort Benning.[82]

Upon his return, Ollie was first assigned to the G-2 section of the 29th Infantry Regiment as an intelligence officer under Colonel Harris Pendelton, Jr., one of the first officers in the new army to have risen from the rank of private. In the autumn of 1928, coaching football again became a part of Ollie's life when he led the First Battalion Musketeers team, and in December, he expanded his repertoire with a turn on the Fort Benning stage as the anxious father in "The Black Suitcase." In January, Captain Reed was appointed the base Athletic and Recreation (A&R) Officer.

Infantry School News (Donovan Library, Ft. Benning)

On June 21, 1929, Ollie was given command of machine gun Company H of the 29th Infantry Regiment. One month later, Company H was featured in a photo spread in the *Infantry School News* for setting the machine gun record in the 29th Infantry.[83] In October 1930, Ollie led eighty-nine men of Company H on an eleven-day march through southern Georgia. The company was recognized for returning to Fort Benning with only two fewer than the number who had started.

"Every man was still in the highest of possible spirit and glad that everyone had made the march. All members of H Company think, and know, they have 'the best outfit in the army.' The work on the hike was carried on like clockwork under the directions of the Company Commander, Captain Reed," the *Infantry School News* noted. Among Ollie's duties during the long march was inspection of the pitched tents. "All that was needed for inspection was a gas mask," the paper quipped in the same story.

Mildred stayed busy hosting and attending social events, as well as acting as secretary of the Chapel Guild and serving in the PTA with Mary Bradley, wife of Major Omar N. Bradley. Mildred was the "room mother" for both sixth and seventh grades at the children's school. Buddy was garnering attention as a twelve-year-old boxer, weighing in at a strapping seventy-one pounds and Ted was becoming a fixture on the school honor roll starting in first grade. Machine gun companies, at this time, were mounted, meaning that Captain Reed also commanded a stable where Ted learned to ride and love horses.

Dinner parties and dances at the Polo Club, serenaded by the 29th Infantry orchestra, helped to make life at Fort Benning pleasant for the whole family. The October *Infantry School News* frequently noted

Ollie's sister, Eva Mae Reed, as a guest at various social events at the base. Eva was seven years younger than Ollie and idolized him. She was wary of Mildred when she married Eva's big brother, but came to love her as a sister. Eva visited from Merchantville where she still lived after moving there to join Ollie and Mildred before they moved to Fort Benning. Twenty-six years old and single, Eva was "being extensively entertained at many lovely parties" during her weeklong stay.

Eva Mae Reed, 1920s
(Courtesy Gladys Moyer)

In February 1931, Captain Reed was sent on Detached Service for two weeks with the Army Air Corps at Maxwell Field in Montgomery, Alabama. An issue of the *Infantry School News* led with an item about Ollie's detail under the headline, "Reed Forgets His Name, or So Aviators Allege," telling of his travails as the brunt of practical jokes at the hands of the aviators. According to the report, Ollie became so flustered upon signing in that he was singled out by Major Walter Reed Weaver, commanding officer of Maxwell Field, intoning for all to hear, "Who is the officer who knows all about himself except his name?"

Two weeks later, Captain Reed's aviation adventure was again a feature in the Benning newspaper, this time in "The Blunderbuss" column:

> Although the science of aviation had developed to a point where flying has lost most of its perils, an Infantry Officer detailed to Maxwell Field for the course in air observation still faces grave dangers which few of his compatriots appreciate.

The visitors from Fort Benning did not actually fly planes during their course at Maxwell, only observed them in flight, but they did have routine duties to perform, such as guard duty. Rounds of the airfield were made on a bicycle with a 32-inch saber by the guard's side. The ride into the unknown could be especially exciting in the mud and darkness. This task got the better of Infantryman Reed, a fact everyone back at Benning would soon know.

> During his recent term at Maxwell Field, Captain Ollie Reed barely escaped a watery grave as a result of his initial midnight solo flight on the ancient two-wheeled relic furnished by the Q.M. Setting forth to inspect his guard with a sabre dangling bravely at his side,

Captain Reed, who hadn't ridden a bicycle in 17 years, suddenly found himself ploughing along a lonely road flanked by water-filled ditches. This mental hazard, as great as it was, would not have sufficed to unhorse the Captain had not his shining Excaliber become entangled in the running gear of his unfamiliar transportation. When this happened, however, the Captain went into a combination nose dive and tail spin which plunged him headlong into one of the ditches while his riderless bicycle landed in the other.

It is authoritatively stated that, as a result of this episode, Captain Reed has made formal application for two weeks back flying pay and membership in the exclusive Caterpillar Club. He stoutly denies, however, that he demanded a parachute for subsequent guard tours.[84]

While Infantry School training was deadly serious, there was often a humorous undercurrent. Mildred remembered a trick her husband pulled during a parade at Fort Benning:

The 29th Infantry had a parade for some distinguished officers. The Generals were really impressed when Pop's machine-gun company passed in review. At the order "Eyes Right," all the mules turned their heads toward the reviewing stand. One of his young officers behind the stand had a mirror that he flashed, which caught the mules' attention. The Generals didn't know that. They thought the animals were extraordinarily well-trained.

The Reed's 1923 Nash Touring car

By the second time around at Fort Benning, the 1923 Nash they purchased after returning from Germany, the car Ollie tried to get Mildred to learn to drive, had worn out. It had a way of its own, one only Ollie understood, making it nearly impossible for anyone else to drive, Mildred wrote:

The old Nash had developed so many idiosyncrasies, no one could drive it but Ollie. A friend took us to Columbus, and we bought a Plymouth that I could manage. I felt sorry for Ollie when men came to take away the Nash. He had to start it for the driver. I think

Love and Sacrifice

there was a tear in his eye, as he watched it disappear. "I feel like I've parted with a faithful horse, or something alive," he mused.

As is often the case with military life, the constant moves and long absences were particularly difficult on children. Both Buddy and Ted had trouble adjusting to their new environs:

> Teddy was bored in kindergarten. We had already done most of the things the teacher had to offer. One day he played hooky. He hid in the garage until time for the afternoon class to begin. I was in the basement doing laundry. Past the high little windows went two pudgy legs. I rushed upstairs in alarm. "Are you sick?" "Yes." "Did the teacher send you home?" "Yes." "Do you want to vomit?" "No." I felt his forehead. No fever. "Well, you'd better go to bed." He was docile, but when Buddy came home from school, he was ready to get up and play. I squelched that idea. He stayed in bed the rest of the day. Maybe I should have let him quit kindergarten. But if I did, he might think he could goof off when he started first grade. Decisions! Decisions!

> Later on, much to our chagrin and distress, we found that Ted had very bad sight. Every year he had had regular school check-ups but we had no idea of any trouble. All this time the poor child couldn't see the blackboard. No wonder school was frustrating! And we considered ourselves good parents! Woe is us!

In first grade at the Fort Benning Children's School, though, Ted made the school's honor roll, in spite of obstacles, and managed to keep up his performance. Outside of school, horseback riding became a favorite pastime. Mildred pointed out the importance of riding in Ted's life:

> At the beginning of third grade he Ted was on the verge of a nervous breakdown. He wasn't assimilating his food so it had to be especially prepared; sun baths, lots of rest and NO school work. I read aloud and we played quiet games. We discovered guppies

Ted's eighth birthday party.

as an entertaining hobby. When he seemed stronger, one of Ollie's corporals took him horseback riding every day.

For Ted's eighth birthday, instead of a party with friends, he wanted a family riding party and a cook-out breakfast in the woods. Ted had become an accomplished equestrian. Mildred recalled that as one of the nicest birthday parties the family ever had.

Sonny Strickland, Ted Reed, and Ted Drewry

Bob Sweet, Bud Reed, Ann Sweet, and Ted Reed

Ted enjoyed playing football with his good friends Sonny Strickland and Jim Drewry. He was even a little sweet on Ann Sweet, daughter of Captain Joseph B. Sweet, an Infantry School instructor. The Reed and Sweet families remained very close for the rest of their lives. The Sweet children called Mildred and Ollie, "Mama and Papa Reed." Bob Sweet was Buddy's best friend for many years.

On May 29, 1931, the *Benning Herald* carried the news that Captain Ollie W. Reed had received orders for new assignments. The Reed family would be returning to Kansas at the Command and General Staff School at Fort Leavenworth.

Ollie was thrilled to be going back to Kansas where he could have a decent newspaper — "the good old Kansas City Star and Times."

As Mildred recalled, "We went to visit kin in western Kansas before reporting to the Fort. On the way driving north, Ted asked what Auntie was like. I said 'Oh, she's a skinny little old gal.' One day at a dinner of assembled relatives there was a lull in the conversation. Ted spoke up 'Mother, I don't think auntie is such a skinny little old gal.' Pop thought that was a rich joke on me. His hearty laugh still rings in my ears. I felt like crawling under the rug."

CHAPTER NINE

Command and General Staff School

School of Application for Infantry and Cavalry, class of 1883. (Combined Arms Research Library)

OLLIE ENTERED FORT BENNING the year before Black Tuesday, October 29, 1929, the day the stock market crashed and America entered the Great Depression. Now, in the spring of 1931, Ollie was returning to Kansas for study at the Command and General Staff School at Fort Leavenworth. Unemployment in the civilian world continued rising and great pressure was being applied to reduce the military budget, but it was a safe, insular life in the army. The pay was low but essential daily living expenses, such as housing, food and medical care, were significantly lower than in civilian life.

Defense budgets remained flat under the ongoing provisions of the 1920 National Defense Act, as did manpower that had been reduced to 150,000 personnel. As the Depression deepened, voices rose for further cuts in funding. The only increase in manpower since the 1920 act was in passage of the Air Corps Act of 1926, increasing the number of airmen to 165,000. Fascination in flying machines had only grown from the days of the romantic flying aces of the First World War, and Lindburgh's Atlantic crossing in 1927 kept attention drawn to the skies, while the infantry, and their budget, languished on the ground.

Furthermore, in 1928, the United States and France drafted the Pact of Paris agreeing to reduce their armed forces to levels necessary only for domestic defense. Three years later, the chief of the Army's War Plans Division, Brigadier General Charles E. Kilbourne, advanced the policy that domestic defense was the primary task of the army for which it had been organized, equipped, and trained. As public opinion, political policy, and the army's own perception of its mission melded in the 1920s and early 1930s, the nation was unprepared for even its own defense. In 1933, Army Chief of Staff General Douglas MacArthur ranked U.S. Army manpower seventeenth in the world.[85] Six years later, as America prepared for another war, MacArthur's successor, General Malin Craig, placed U.S. force strength at eighteenth.[86]

Into this changing martial landscape, Captain Reed entered the graduate school for army officers - the Command and General Staff School (CGSS). The focus of the two-year CGSS program was the development a cadre of officers for general staff positions and command of divisions, corps, and armies. The recommendation for advancement to the War College and General Staff were the brass rings for aspiring officers.

Fort Leavenworth was established in 1827 as a forward operating base protecting the northern neck of the Santa Fe Trail. It was named in honor of Colonel Henry Leavenworth, founder of the frontier outpost, known for his exploits in the War of 1812, and in battles against the Plains Indians. CGSS started as the School of Application for Infantry and Cavalry in 1881 and graduated its first class in 1883. Captain Reed was entering the 50th anniversary class.

It was a distinct honor to be chosen for the CGSS at Fort Leavenworth,[87] a school of advanced study for select members of the officer corps, and a significant step up the promotion ladder. Ollie was recommended by the office of the Chief of Infantry, Major General Stephen O. Fuqua, signed by Colonel Lorenzo D. Gasser for the chief, to be among the 130 selected for the 1931-33 term.[88] Competition was stiff as only ten percent of the total officer corps qualified for CGSS, primarily among the ranks of captains and majors. The "hump" in the officer corps, created in the build-up prior to World War I, had matured into a bubble at the higher ranks. In April 1917, there were

5,960 officers in the Regular Army. By the Armistice in 1918, there were 203,786. The majority of the post-war drawdown in forces concentrated on conscripts and reserve officers, while the Army sought to retain the best of the Regular Army officer corps despite Congressional efforts to reduce them.

In their 1977 survey of the Official Army Registers from 1925, 1933, and 1940, Edward Coffman and Peter Herrly found that more than a third of all officers came from the Midwest, while slightly fewer were from the South. Almost a third rose from the enlisted ranks, most entering during mobilizations for the Spanish-American War or the First World War. More than thirty-seven percent of officers were West Point graduates, as were more than seventy percent of generals.[89]

Promotions slowed to a crawl due to the mobilization hump and budget cuts targeting reduction in personnel. During the interwar years, it took three years to reach the rank of first lieutenant from second, another seven years to captain, seven more to reach major, another five years to lieutenant colonel, and six more to full colonel. Budgets were cut, pay was lousy, and morale was low, but what waited for them outside the service was the Great Depression. Pay grades had risen only eleven percent since Congress established new levels in 1922. Generals earned only between $6000 and $8000 per year but officer wages still looked good to the general public. In fact, between 1928 and 1933, army pay had been cut fifteen percent plus one unpaid month as part of the New Deal programs. The stress on family budgets was eased somewhat by lower costs for the essentials of housing, food, clothing and medical expenses.[90]

From 1929 through the early 1930s, Kansas was generally spared the ravages of the Great Depression, particularly when compared with more urban states around the country. Kansas was, and still is, an agricultural state. Farmers were all too familiar with hard times and they held tough through adversities, such as those brought on by weather. Plus, wet conditions prevailed through the 1920s and the introduction of hard winter wheat into the growing cycle helped build profits and wealth. With this, more and more land was being tilled for ever-larger wheat crops, something that would come back to haunt the region in the next decade. Areas of Kansas that relied on mining felt the sting of the Depression as commodity prices dropped. Coal mining in eastern Kansas had already peaked during the war years, as did lead, zinc, silver and cadmium in the 1920s, all falling during the Depression. The Carey Salt Mine, today the Hutchinson Salt Company, processed rock salt from underground caverns mining ancient salt seas that once covered the landscape. Salt was the exception among mined materials in Kansas as rock salt was used in icing down rail cars carrying produce and other perishables before refrigeration. Retailers in small towns near larger urban areas began losing shoppers with the rise in automobile use and improved roads as people searched for better prices and availability. In all, Kansas fared well, until the rains stopped and the winds swept the dried earth from the landscape into the air.

A student-officer's performance at the end of the two-year program had a direct impact on their promotion potential and academic trajectory continuing at the Army War College. While the Company Officers' Course at the Infantry School, when Ollie attended in 1923-24, focused on command at the company level, primarily overseeing rifle and weapons platoons, with an introduction to commanding a battalion, the two-year CGSS program took training to a new, intense, and exhausting level, developing the tactical and strategic skills necessary to command a corps, division, or a field army. Students spent every waking hour studying maps and manuals, as well as finding solutions to battlefield situations on a large scale. This was a key part of the army's new emphasis on education and its "applicatory" method of learning through individual practical applications. Officers who graduated successfully from CGSS would assume command with responsibilities for thousands of lives, millions of dollars in equipment and supplies, and the conduct of a military operation that could alter the outcome of a war.

The time spent at Fort Leavenworth was among the toughest two years in an officer's career. Since Captain Reed was not a graduate of a service academy, or even college, he had to work exponentially harder than virtually everyone else to succeed. Finishing the program below average meant an end to one's career advancement. Class standing at the end held great significance.

A 1950 history of the program provided an overview of the CGSS program:

> Education of the professional soldier to meet the demands of future conflict must be continuous, just as it is in any other profession, beginning with basic instruction in the military art and culminating in the study of doctrines and techniques at a post-graduate level. Patriotism and valor alone will not make leaders of our future Armies. Professional officers must study constantly. The student of higher military education must be intelligent, diligent, and mature; he must have a firm grasp of basic tactics and technique and be seasoned with field experience.
>
> Thorough military education combined with the personal qualities of leadership provides intelligent leadership in combat.[91]

General William Tecumseh Sherman formulated the foundation of the school's experiential curriculum in 1881 under the original name – School of Application for Cavalry and Infantry.[92] Beside his service in the military, Sherman had been the first superintendent of the Louisiana State Seminary of Learning & Military Academy in Pineville, Louisiana (today, Louisiana State University). As an 1840 graduate of West Point, entering the academy at the age of sixteen, Sherman recognized the value of ongoing and evolving advanced education. After the Civil War, Sherman, along with fellow generals Grant and Sheridan, sought a new way to reorganize the army after witnessing numerous deficiencies in tactical leadership and confusion in the command ranks during combat. They recognized advanced education as the solution.[93] Just as Sherman directed the school to teach the lessons learned from the Civil War, adapting to new tactics and technology, the same held true in the interwar years when military education was turning away from the tactics - stabilized, positional, and trench warfare - employed in First World War and returning to the offensive spirit of open warfare.[94]

In an address to the Army War College in 1924, General John J. Pershing had high praise for these military scholars:

> During the World War, the graduates of Leavenworth and the War College held the most responsible positions in our armies, and I should like to make it of record, that, in my opinion, had it not

been for the able and loyal assistance of the officers trained at these schools, the tremendous problems of combat, supply, and transportation could not have been solved.[95]

In his 1970 *Combat Commander: Autobiography of a Soldier*, retired General Ernest Nason Harmon, 1933 CGSS graduate and West Point alumnus, wrote about his two years at Fort Leavenworth:

> The two years I spent at Leavenworth were the most difficult years of my training. About 250 officers matriculated, and some of us had almost lost the habit of study; we were warned the class would be thinned down to 125 for the second year. I was able to make the second year, but only with my wife's encouragement and practical management. There were now five children in a crowded house, five children afflicted with the colics, fevers, and ailments of the very young. I studied upstairs and downstairs, often far past midnight, and my disposition at home became as mean as that of a starving prairie wolf, or - as one of my friends suggested - a cobra without a convenient snake charmer." [96]

Harmon commanded the 1st Armored Division in World War II, and gained renown for regrouping the Army II Corps after it faltered during the battles at the Kasserine Pass in North Africa.

In 1931, with a wife and two growing boys, ages nine and twelve, Captain Reed entered the grueling CGSS course. Being one of the youngest in the class (the average age was forty-five and he was thirty-four), Ollie, with barely his two years at Kansas State, was in a class of 125 that included twenty-six West Point graduates, most from the classes of 1915–17. The class of 1915 was referred to as "the class the stars fell on," as fifty-nine of its members went on to attain the rank of general. Ollie's class was evenly divided by the source of commission: one third were graduates of West Point, another third were commissioned after college graduation, and the final third, like Captain Reed, had risen through the ranks. A rating of "excellent" was required from the chief of Infantry for Ollie's admission to the program as a CGSS appointment was an acknowledgment of each appointee's military career success and professionalism, with bigger and better yet to come. Over the course of World War II, a CGSS graduate commanded every one of the ninety divisions and twenty-four corps.

Love and Sacrifice

CGSS class of 1933, Captain Reed is at the lower left corner. (Combined Arms Research Library, Ft. Leavenworth)

The class was comprised of fifty-one members of the Infantry, seventeen from the field artillery, eleven from the cavalry, ten each from coastal artillery and the air corps, five engineers, and three from the signal corps. Another twenty came, ten each, from the general services and upon the recommendation of the secretary of War.[97]

The arduous classroom work proved difficult for families, and it was emphasized to incoming students, especially their family members, that the road ahead would be difficult and success could only be realized with the loving support, as well as patience, of wives and children.

Fort Leavenworth teemed with children. The Reeds lived on the ground floor in Apartment C at 315 Pope Avenue, known as Gregg Hall in honor of Captain John C. Gregg, a member of the class of 1897, Infantry and Cavalry School, who was killed in action in the Philippines in 1899. Their apartment adjoined three others in the red brick building, with another four units upstairs. A common hallway connected the apartments. Infantry Captain Charles H. Karlstad, who attended Infantry School with Ollie, lived in unit D with his wife, Barbara, and ten-year-old daughter Celeste. Upstairs, Captain Donald A. Stroh lived with his wife, Imogene, and their two children, Imogene and Harry, who were the same ages as Buddy and Ted. Bill

Ted Reed, Mary Almquist, and Bill Pendergrast at Ft. Leavenworth

Pendergrast and Mary Almquist, Ted's good friends, lived in units F and G, respectively, with their fathers, Captain Alan Pendergrast, and Captain Elmer H. Almquist. Captain Louis Cansler's two children, Virginia and Rogers, also lived upstairs, in unit H. Marine Corps Captain Samuel C. Cumming, who had been awarded five Silver Stars in World War I and would later earn a Legion of Merit in World War II as a Marine Corps colonel (later a brigadier general), lived in unit A with his wife, Eula, and four-year-old son, Sam Junior. Captain Stonewall Jackson, an instructor from the Infantry School, and his wife, Dorothy, were the only couple without children in the building. The Reeds' best friends, the Sweet family, lived next door.

Ollie and Mildred's apartment had a sunroom where Mildred enjoyed spending time reading and sewing. Two identical rows of six New York-style apartment buildings, as they were called, backed one another with a common courtyard in between. The front row, where the Reeds lived, faced Pope Avenue and the center of the base. The back row overlooked the Grant Avenue Polo Field (across Doniphan Avenue), surrounded by the golf course. This was a land of wide-open spaces and adventure for two growing boys and a slew of other children with whom to share the fun. The YMCA, a large red brick building, was located directly across Pope Avenue from their apartment. Dances, bowling, swimming were always underway. The Post Exchange (PX) was located at the end of their block. Clocks meant little to the children growing up on military posts as daily life revolved around the series of post bugle calls.

Ted was enrolled at the Post School, located next to the Post Chapel on Scott Avenue, and Buddy started as a freshman at Immaculata High School in the town of Leavenworth.

Ollie would bring some of his studies home to Buddy and Ted, Mildred remembered. "It was a delight to us to introduce our boys to the fascinating world of books, and to take them to historical places. I vividly recall Pop and the boys in the sand-pile laying out the terrain for Custer's Last Stand. They had soldiers and Indians made of lead. Pop explained each move and showed them what might have been a more advantageous procedure."

The legends of the frontier fort helped bring history to life. One such character was General George Custer, a well-known figure at Fort Leavenworth, with a plaque in his honor in the Post Chapel and a ghost story in his name. He is said to haunt the place where he was found guilty of abandoning his command in a court martial in 1867.

Most comforting to Mildred was the Post Chapel, built in 1887 using stone quarried from the fort's grounds and labor from the Leavenworth prison. Reverend Luther D. Miller was chaplain at Fort Leavenworth from 1928 to 1937. After graduating from the Chicago Theological Seminary, he served as an army chaplain in Tientsin, China. Reverend Miller, with the approval of Lieutenant Colonel George Marshall, repaired the chapel and established a service club featuring both a library and gambling room to keep soldiers out of trouble with the myriad local vices. Reverend Miller was never present while gambling was underway, but soldiers knew the slot in the center of the table was meant for money to "sweeten the kitty for the church."[98]

Rev. Luther D. Miller at the Post Chapel

Ted and Margaret Berry

Ted was an acolyte at the chapel and took regular music lessons with the church organist, Margaret Berry, who provided music for chapel services for over fifty years. He came home one day and announced that he was going to confirmation class to prepare for church membership. Bud joined him and Reverend Miller confirmed the boys at the baptismal font in the Post Chapel.

Reverend Miller rose through the ranks, becoming Army Chief of Chaplains as a major general. After retiring from the army in

1949, Reverend Miller became Canon at the National Cathedral in Washington, DC. He was followed into the ministry by his son, Luther Miller, Jr., who served in the infantry in the Pacific during the war. The younger Reverend Miller became pastor of Saint David's, a small Episcopal church in Northwest Washington, DC, where Mildred and Ted saw him years later. Without thinking, Mildred immediately addressed Reverend Miller as he greeted them at the church door with his childhood nickname, "Chappie." Mildred recalled standing there with the two grown men, still thinking of them as children.

"Distinguished men, though they were, they were still little boys to their mothers."

The class of 1933 was a mix of officers who had served in World War I and those who had not. Alexander Barnes, author of *In a Strange Land: The American Occupation of Germany, 1918 – 1923*, said that the animosity that might have existed between the two groups when they served in the post-war American Forces in Germany (AFG) had all but dissolved by this time.

"It doesn't appear there was any real overt discrimination against officers who didn't fight in France - Bradley, Eisenhower and Lightning Joe Collins all had pretty successful careers - I'm sure there was some animosity but the interwar army was so small, they all seemed to know each other and let personalities determine friendship rather than service in the trenches."[99]

The Great War provided essential lessons for these future leaders, who would face another European conflict in less than a decade. Although he had not served in combat, Captain Reed's major theses examined the "Capture of Jerusalem and winter operations in 1917–1918," a group research study detailing the movements of British forces during the Palestine campaign against the Turkish Army, and "Critical analysis of Turkish military operations during the Palestine Campaign, 1917–1918," an individual research study by Reed describing Turkish military operations, including descriptions of political and military conditions prior to the campaign. Ollie had no personal experience with the region or topic, but CGSS students studied past conflicts for lessons to be learned in future planning and implementation.

Another classmate, Major Charles H. Gerhardt, a World War I

veteran, wrote about the 89th Division in the Meuse-Argonne, a unit with which he had served as a second lieutenant in the Cavalry. Later, Major General Gerhardt assumed command of the 29th Infantry Division in 1943, following his command of the 91st Division.

Teaching accomplished officers was not an easy task for CGSS instructors, and it was not uncommon for students with expertise in certain areas to challenge them, but, just as it was for Plebes at West Point, know-it-all student-officers could suffer consequences with deficient marks. Competition for good grades affected an individual officer's standing for future promotions, and this provided a good incentive for doing one's best. Too many deficient marks could have a significantly negative impact.

Classification	Academic Rating	Name	Recommended for Higher Training In:				Recommended For:	Considered Especially Qualified For:								
			Command	Staff Duty	G. S. Eligible List	Course at A. W. C.		Division				Corps				
								G-1 8	G-2 9	G-3 10	G-4 11	G-1 12	G-2 13	G-3 14	G-4 15	G-4 17
1	2	3	4	5	6	7										
G	EX	Reed, Ollie W., Captain, Infantry	No	Yes	Yes	Yes		Yes	Yes	Yes	Yes	Yes	Yes	Yes	Yes	Yes

Captain Ollie Reed's final evaluation from the CGSS program. (Combined Arms research Library)

Completing the CGSS course was not an automatic entrée to further study at the War College or placement on the General Staff Eligible List – those had to be earned through hard work and good grades. The final rank-order of the graduates, based upon academic grades, was critically important in an officer's career. The school reported each student's order of merit to the War Department with a recommendation as either "qualified" or "unqualified" to continue their education as a general staff officer. While field observations of the officer-students in action were made by a handful of instructors, a board of twelve to fifteen experienced officers made CGSS evaluations and recommendations over time and through a range of trials.[100] Such competition was essential to Army life and particularly, in an officer's career advancement. Out of 125 graduates, Captain Reed placed forty-seventh with an Excellent Academic Rating. Common sense and practical application of lessons learned meant much more than an educational pedigree in the CGSS program. Ollie also received a recommendation for inclusion on both the General Staff Eligible List and for further study at the Army War College, and was listed as qualified to serve on the general staffs at both the division and corps levels.[101] He also was among only six graduates to receive an overall

Excellent, rating but the one box that was not checked for further training and advancement was Command. While Captain Reed was a highly qualified and universally respected officer, there were questions about his ability to make tough decisions where men serving under him would be sent into the line of fire and face possible death.

In spite of the intense competition to gain one of the approximately seventy-five positions in the War College, Ollie opted not to follow through with the opportunity. Instead, he accepted assignment as professor of Military Science and Tactics at Wentworth Military Academy, just a short distance east along the Missouri River from Fort Leavenworth in Lexington, Missouri. A posting as an instructor and head of an ROTC program was considered within the officer corps as a prime assignment in the interwar army.[102]

More important than the prestige of the teaching assignment was the opportunity it presented for realizing his long-range dream of an appointment to West Point for Buddy. Ollie kept an eye on the advancement of officers who were West Point graduates and knew full well that he made a mistake in turning down his appointment eighteen years earlier. Ollie had made his mistakes, learned from them, and would try to make certain that his son would not repeat them.

The War College could wait.

Grant Hall and Bell Tower,
the HQ of the CGSS in 1933.

CHAPTER TEN

Back to School, for Bud's Sake

Cadet Reed and his father at Wentworth.

FIELDS OF CORNSTALKS, harvested the month before, withered in the Kansas heat. The browning remains stood in regular rows, like soldiers standing at attention, too long in the sun as their arms sagged by their sides. Ollie told the boys that corn used to grow twice as tall in years past, but without rain, they didn't stand a chance. Frequent blasts of dust-filled wind blew across the highway from the north. The first major dust storm of the Dust Bowl era was three months away, but already there were signs of worsening conditions.

The three hundred miles west to Norton and back again, plus another sixty-five miles to Lexington, made for a very slow drive in 1933. Although U.S. Route 36 was a brand-new interstate highway, most of its surface was packed gravel. Whether gravel or paved, speed limits at this time held cars to forty miles per hour outside of cities or villages, and twelve miles per hour inside city or village boundaries.[103] Luckily, they had the summer months to visit family and then get fourteen-year-old Buddy and eleven-year-old Ted ready for their new schools in their new hometown as summer was giving way to autumn.

The bluffs rising from the Missouri River were a welcomed sight in the distance after so many hours and days of largely flat, endless terrain. The Reeds lived in six different places over fourteen years and hoped, as they boys were entering their teenage years, they could settle down in Lexington for some "growing time."

Crossing the Missouri River on State Route 13, over the Lafayette-Ray County Bridge, the road ended abruptly in a T. There, the Reeds faced a whitewashed block building jutting from the base of the escarpment, bearing the inscription, "Memorial to World War Veterans of Lafayette County, 1925."

Lexington's Madonna of the Trail
(Library of Congress)

Turning left, the Reeds wound up the face of the bluff until they were greeted by one of twelve Madonna of the Trail statues dotting the Old National Trails Road from Bethesda, Maryland, to Upland, California. The Daughters of the American Revolution erected the identical eighteen-foot-tall monuments honoring pioneer women across the country, in 1928 and 1929. Each monument depicts an identical woman dressed in traditional garb, holding a shotgun in her right hand and an infant in her left, while a small child tugs at her dress. The Lexington Madonna had been dedicated in 1928 by the presiding judge of Jackson County, Missouri, and president of the National Old Trails Association, Harry S. Truman, the future president of the United States.

Lexington, Missouri is a picturesque small town of antebellum homes and Greek Revival public buildings perched above the Missouri River, about forty miles east of Kansas City. Much of the business district of Lexington was built during the latter years of the nineteenth and early twentieth centuries when Lexington thrived as a commercial center. Many of the names of Lexington residents have remained the same through the decades, starting with the Aull family who helped establish Lexington as a mercantile center along the Missouri. By the mid-nineteenth century, steamboats had replaced wagon trains bringing settlers to Lexington, which continued to grow at the head of the Santa Fe Trail. The Missouri Compromise, passed by Congress in two parts during 1820, brought most of the Missouri territory into the Union, but the "compromise" part of the legislation allowed Missouri to remain a slave-holding state. The 1854 Kansas-Nebraska Act left it up to Missouri's neighbors whether they would be free or slave. Western Missouri, bordering all of eastern Kansas, was a slave-holding stronghold with as many as fifty thousand slaves in bondage. Kansas, with an influx of easterners looking for free land, was leaning toward abolition. The hostilities broke into violence in 1856 when the abolitionist town of Lawrence, Kansas, was attacked by a band of several hundred "Border Ruffians" from Missouri ransacked and burned parts of the town. Two days later, John Brown led a band of Kansas abolitionist raiders in an attack upon settlers along the Pottawatomie Creek in Missouri, killing five. The Pottawatomie Massacre ignited a series of retaliatory cross-border attacks that would be known as Bleeding Kansas, regarded by many as the true beginning of the American Civil War.

On June 11, 1861, Brigadier General Nathaniel Lyon met with secessionist Missouri Governor Claiborne Fox Jackson and Major General Sterling Price, head of the Missouri State Guard, at the Planters' House Hotel in St. Louis. He was looking to reach an amicable agreement to avoid bloodshed. Instead, the conference ended after four hours with Lyon famously declaring to Jackson, "This means war."

A cannonball from the September 1861 Battle of Lexington remains embedded in a pillar of the Lafayette County Courthouse in Lexington. Remembered as the "battle of the hemp bales," secessionist forces soaked bales of hemp in the Missouri River overnight to prevent cannon fire from penetrating them the next day.

Following the Civil War, the completion of nationwide rail networks, with Kansas City as a major hub, curtailed steamboat transport, diminishing the need for coal from the Lexington mines, and removing Lexington from its transportation and supply prominence. Once calling itself the "City of the Future," Lexington faded from its glory days. Brewing, banking, and retail became the major industries.

At the turn of the century, Lexington was known as the "Athens of the West," hosting several institutions of higher learning, including Elizabeth Aull Seminary, Lexington Ladies College, Central College for Women (formerly the Masonic College), and the Wentworth Military Academy. By the 1930s, only Wentworth was still in operation, and only barely. Just as the town had done to make payroll, Wentworth also borrowed against assets to keep operating.

Wentworth was founded in 1879 by Benjamin Lewis Hobson as Hobson's Select School for Boys, in the First Presbyterian Church of Lexington, where his father was pastor. The schools in Lexington exclusively catered to young women, and Hobson wanted to fill the educational void for boys. Later that same year, local banker Stephen G. Wentworth, mourning the loss of his son, purchased the school and renamed it the Wentworth Male Academy in his son's honor. In 1880, Sandford Sellers became the first principal, superintendent and president of the Wentworth Military Academy, and remained active in the school until 1938 when he died at the age of 84.[104] Since then, Wentworth has been under the continuous guidance of the Sellers family, including the current president, William Wentworth Sellers, the great-grandson of Sandford Sellers and a maternal descendant of Stephen Wentworth.

Under the National Defense Act of 1916, Wentworth was one of only seven military schools authorized by the War Department to organize a senior division of the SATC. With the U.S. edging toward the World War, the school rapidly grew from 155 students in 1915 to 377 enrollees in the 1917–18 school year.

By 1933, the year the Reed family arrived, the Wentworth Military Academy had fallen on hard times. The Depression and demilitarization of the interwar years contributed to the school's declining enrollment, bottoming out at 140 students. Wentworth was forced to borrow millions on the value of land holdings just to stay afloat.

Into these dire straits, Captain Ollie W. Reed reported for duty as professor of Military Science and Tactics on September 9, 1933. The school opened three days later with 143 enrolled students.

Such circumstances were not unfamiliar to Ollie, having resuscitated the Military Science and ROTC programs at Drexel in Philadelphia during the 1920s. Ollie got to work right away. Captain Reed's cadets had their first public appearance, joining the local American Legion, in a November 1933 Armistice Day ceremony, just two months after his arrival. Training was constant; even when they traveled to the Kemper-Wentworth football game, Captain Reed made certain they used the occasion for training in organizing a troop movement by rail. Student officers were under careful observation for the control of their assigned units. Reed established traditions at the school, such as the awarding of a .45mm pistol and officer's saber at commencement to the leading graduates. In his second year, the ROTC band gave a full-length public concert just days after classes started, and little more than two weeks later, the ROTC Battalion staged a dress "Welcome Day" parade through downtown Lexington with five thousand people lining the sidewalks.

Wholesale changes were underway at Wentworth, as Colonel James M. Sellers, the son of Sandford Sellers and a highly decorated World War I Marine Corps veteran, became superintendent of the school. He recruited Lester B. Wikoff as treasurer and business manager to help revive the business side of the school. Gradually, the school began to recover, as noted in a history of Wentworth:

> The heavy task of rehabilitating the tottering academy, paying off old debts, restoring financial credit and building up the corps of cadets fell to Colonels J.M. Sellers and L.B. Wikoff…By the opening of the 1935-36 session the situation had been materially eased. One of the main factors in the academy's ability to survive the harsh years of adversity was the Junior College – enrollment

increased every year in the period.[105]

By 1937, the rush to registration surprised and overwhelmed school officials on the opening day, Sunday September 12, when 197 students enrolled, rising from 172 the year before. "Cars from every state lined the streets surrounding the academy," the local paper heralded. Fortunately, Major Reed had already prepared "B" barracks for opening after being closed due to falling enrollment.

High school football games were big events in town. The Lexington Junior-Senior High School played at the adjacent Goose Pond field while Wentworth hosted their games at Memorial Stadium. Locals lined the streets to watch the school bands and pep squads march in advance of big games.

In town, kids could catch a matinée at the Mainstreet Theatre for a dime and then go next door to have a five-cent hamburger at the Snappy Service, or an ice cream in the soda shop owned by the Anton family that was in the same building as the theater.

A 1936 Mainstreet Theatre marquee displayed "The Great Ziegfield starring William Powell and Myrna Loy, playing Sunday, Monday and Tuesday," with teasers underneath with the come-on: "Un-cut – Unchanged – Three hours of thrills, songs, beautiful girls, stars and heartthrobs exactly as played for five solid months at the Astor Theatre in New York." The Eagle Theatre was a small movie house where Gene Autry and Buck Jones westerns regularly filled the two hundred seats.

For adults, there was dancing at the Dreamland Ballroom on Saturday and Sunday nights. For ten cents, couples could dance the night away to the sounds of Charles Stephaniak and his Sophisticated Swingers. There was also the Dreamland Grill, where hot dogs and hamburgers cost five cents and ice cream sodas ten, and Dreamland Lanes was Lexington's bowling alley. The grill and bowling alley occupied the first floor of the two-story building on Franklin Street, and the ballroom took up the entire second floor.

Lexington had two distinct faces. By day, in spite of harsh economic times, it was a bustling mercantile town where shoppers came from all around the region to buy their wares. Nighttime was realm to a myriad of saloons, particularly along Main Street, which gained the nickname, "Block 42," for the forty-two saloons lining the street (some topped with gambling or prostitution upstairs). Among them were the Modern Cafe, owned by Jimmy and Gussie Lorantos, and Pat's Bar, where Hollywood western movie star, Tom Mix, rode his horse, Tony, into the saloon for a fifteen-cent beer. Riley's, founded in the 1890s by Irish immigrant Forrest Riley, was housed in the "keyhole building" for its distinctive keyhole shaped stained glass window above the entrance.[106] Prohibition had done little to daunt Block 42. Local police tended to turn a blind eye, much to the frustration of federal revenue agents.

On December 5, 1933, the Twenty-first Amendment repealed Prohibition, but Lexington was slow to legitimize saloons and the possession of alcoholic beverages. It took four months for the first liquor licenses to be issued in the town, two of them to druggists (drug stores were typical fronts for bottled liquor sales during Prohibition) and the third to Cue's Buffet at Tenth and Main. Money was still made from a lively bootlegging trade well after Prohibition ended. In January 1935, federal officers from Kansas City, without informing local police, conducted three raids in Lexington, arresting fourteen men, including Bob Kirkpatrick and Ben Simmons for their 150-gallon still and 1700 gallons of mash. Missouri state liquor control officer E.J. Becker launched a drive in Lexington to get rid of saloons that were not paying taxes. Prosecuting Attorney Arch Skelton declared that all slot machines be banished and all games of chance, such as drawings, raffles, and card nights, cease in Lexington.

Meanwhile, Kansas City gangsters operated with relative freedom under the corrupt regime of political machine boss Tom Pendergast. Frequently, bodies that had been dumped in the Missouri River at Kansas City surfaced near Lexington. The results of feuds or outright killings might be left lying along the countryside roads or even on the Lexington Bridge. Pendergast represented both the best and the worst of these times. Kansas City's jazz age blossomed during his regime, as did the architectural splendor that was funded with public works money benefitting the Pendergast machine and their concrete business. Speakeasies and gambling halls also flourished. Harry Truman was a Pendergast protégé, earning the boss's support in his rise from county judge to the U.S. Senate.

Entine's Department Store anchored the corner of Main and Ninth, and the Palace of Sweets doubled as the local bus station, where Missouri Pacific and Greyhound buses serviced passengers going to and from Chicago, St. Louis, and Kansas City. The local pool hall served no beer or liquor, except at the card game upstairs, and was forbidden territory to many Lexington youth. The Lexington Night Club opened at the corner of Ninth and Main as soon as Prohibition ended. The nightclub is best remembered for bringing in a Kansas City band in the 1930s, led by a young man named Count Basie.[107] Even after Prohibition was repealed, bootleggers remained active in the area, dodging taxes and operating their stills on sandbars in the river, giving name to such places as Bootleggers Island.[108]

The Great Depression, beginning with the stock market crash of 1929, wrought hardship upon the town, as it did across the country, pushing Lexington to the brink. The once-prosperous coalmines that serviced the riverboats and continued operating after their demise closed. Families lived on houseboats floating on the Missouri and Mississippi Rivers much like in the days of Mark Twain.

With the March 1933 inauguration of President Franklin Delano Roosevelt and his New Deal programs, hope slowly began to creep back into the American spirit over the coming years. Still, many people in Lexington were not as fortunate as Ollie with his army salary. The year before the Reeds arrived, Lexington had been unable to pay teachers' salaries, and the city sold bonds on property it owned just to make payroll. Unemployment remained well above twenty percent.

In early 1934, nearly ten percent of Lexington residents were receiving government food aid. On the first day a soup kitchen opened at the local school in December 1934, eighty children lined up for a hot meal. By early 1935, the kitchen was feeding nearly 140 children daily. The town established a 190-acre community garden with a canning operation to help feed the hungry.

Gas station hold-ups were frequently reported in the local newspapers. Petty crime was rampant – chickens, pigs, and other livestock were stolen from farms, and even overcoats were stolen from churches during services. In February 1935, three bicycles were stolen, including Teddy Reed's.

Suicides in Lafayette County were commonplace, with individuals taking their lives by gunshot, hanging, poisoning, and drowning. In November 1936, fifty-five-year-old May Hutton drowned herself in a well over her failing health. Accidental shootings sometime bordered on the bizarre, such as when a teenager died by accidentally shooting himself in the head when trying to brush away a spider's web with his shotgun while hunting.

Weather played a heavy hand against the people of the Midwest during these years, as well. Summer heat consistently stayed above one hundred degrees without rain. On July 20, 1934, when the temperature reached 109 degrees and the average for the month was 102, old timers were quoted as saying that it was "the awfullest heat" they could remember since 1901. Drought and wind brought dust storms so thick they blocked out the sun, followed by torrential rainstorms accompanied by high, sometime cyclonic winds and lightning. The country had lost five million acres of tillable soil, transformed into a wasteland, over just two years, much of it the soil tilled in Kansas for the boom in winter wheat. April 14, 1935, became known as "Black Sunday" as "black blizzards" of topsoil swept over the Great Plains.

The July 17, 1936, *Lexington Intelligencer* ran the hopeful headline, "Relief Forecast by Moon Gazers in Chicago for Middle West Agricultural States Soon," as vast migrations of grasshoppers appeared as opaque clouds against the sun.[109] Heat continued into autumn, delaying the start of the 1936-37 school year, and winter brought long stretches of below-zero temperatures, blizzards, and ice storms.

A November 1936 *Lexington Intelligencer* headline declared that the Depression was passing in the eastern U.S., with unemployment down and wages climbing, but adjoining the story was a photo illustrating an army of vagrants invading the South and the West, with a caption telling of Florida and California's newly established border patrols to turn back the thousands of drifters who were heading their way during the colder months of the year.

Buddy faced a tumultuous start at Wentworth. He completed his freshman year at Immaculata High School in Leavenworth. Now, the second-year cadet was no different from the other Wentworth

Ted, second from left, and Buddy, right, wearing his Wentworth sweater.

cadets, living on campus, and going through the rigorous daily routine. What set him apart was his life as an army brat and, worst of all, being Captain Reed's son. The situation was awkward as Bud had not lived in one place long enough to have the confidence to quickly make friends, so he had difficulties adjusting. Over time, he developed leadership skills and a steady stream of promotions. His father viewed Wentworth as Buddy's path to West Point. Ollie had seen MacArthur, Bradley, and Eisenhower advance but knew the limits of his own promotion potential. He wanted better for his son, and Buddy would not disappoint.

Wentworth morning drill before Captain Reed.

The typical day for a Wentworth cadet was full, from reveille to taps, on the theory that keeping a cadet constantly busy, without exhausting him on a single task, helped develop a work ethic critical to success in future military and civilian life.

Reveille blew at 6:40 a.m. each morning, and cadets had twenty minutes to get washed and be in the dining hall for a breakfast lasting only twenty-five minutes. Afterward, they had thirty minutes to prepare their rooms for daily inspection. They had three, fifty-minute classes each weekday morning, followed by an hour of intensive military drills at 11:00 a.m. under the guidance of Captain Reed.

On some days, the routine could be unexpectedly broken, like on November 3, 1936. Cadets were routinely marching along a road near the campus as part of their morning drill, when suddenly seven Grumman fighter planes dove from the sky and "attacked" the startled cadets as they marched in a column along a winding road two miles north of the campus. Ollie had arranged for the surprise attack by Marine Corps Aviation Unit from Fairfax Airport in Kansas City.[110]

Wentworth morning drill.

Otherwise, each day consisted of an hour-long break for lunch and time for the cadets to relax, and then it was off to the chapel for what was considered a "morale builder," starting with a song and a short religious exercise led by Wentworth Chaplain Reverend Glenn Maxwell, followed by announcements, pep speeches, musical numbers by members of the corps and outside talent, and topical lectures. Sometimes Bud would take this opportunity to play the piano for his fellow cadets, "Anchors Away" being his favorite song.

The afternoon led off with three additional academic class periods, and then it was time for physical fitness education and sports teams. Each cadet was expected to participate in athletics, whether at the company or varsity level, in the belief that physical fitness helped develop mental fitness, as well as teamwork – all critical facets of success in military and civilian life.

At 5:50 p.m., Recall sounded and cadets rushed to shower and prepare for dinner precisely at 6:25 p.m. Their evening meal was followed by an hour of freedom, a two-hour study period where cadets could remain in their own room or go to a study hall, and an final forty-five minutes for further study or reading. Finally, Taps was blown for cadets to be in bed at ten o'clock. Junior college cadets were allowed the privilege of added time before lights-out.

With good behavior, Wentworth cadets could earn the privilege to leave campus on weekends and go into town for activities, such as movies and meals. The young cadets walked along the Lexington sidewalks in their neat grey uniforms, never permitted to wear casual clothing off campus, garnering great respect among the local populace. The handsome young men drew the special attention of the town's girls, who would follow them, giggling. It was common for Lexington area boys to finish high school and then enter the Wentworth junior college.

Buddy's Wentworth photograph, inscribed, "Your Cadet," for his mother.

Bud achieved the rank of lieutenant of cadets in his high school senior year, and in June 1936 received the Major Henry Fox Award, a saber, in recognition of the best platoon. In November 1936, he was promoted to cadet captain and given command of C Company. At that month's homecoming, highlighted by a game against the Kemper Military School, C Company was presented the cup as the winning barracks at the homecoming dance for their decoration of the building exterior. They had covered it with streamers in the school colors of red and white, hanging signs proclaiming, "Beat Kemper" and "Welcome Old Boys." The annual game pitting Wentworth against rival Kemper Military Academy was dubbed "the little Army-Navy game," bringing hundreds of cadets and fans to one another's town for days of festivities.[111] Corps of cadets from both schools, along with their marching bands, paraded through the streets of Lexington and Boonville, Missouri. The Kemper band, dressed in white leggings, formal uniforms, cross stripes and white caps, were led by a drum major and two baton twirlers. There was a bonfire the night before the games, highlighted by the appearance of eighty-six "Old Boys," former graduates of Wentworth.

Although Buddy shared a love of football with his father, standing at little over five feet five and weighing only 136 pounds by the time he left Wentworth, Buddy was never happy with his short stature and had to be content with playing company football. In his senior year he did receive a varsity letter, even though he did not play a minute with the top team. Still, he was recognized on the 1936 Wentworth Homecoming roster: "Ollie W. 'Buddy' Reed is an ex C Company player and a good one at that. This is his first year of varsity competition and so far, though he hasn't played much, he has proven to be a valuable assistant to the team." A head shorter than the player standing next to him in the team photo, Bud proudly wore the number seven.

He once told his mother that he wished he had grown into a big fellow like his father or brother. To this, she told him he should have picked a bigger mother. He affectionately responded that he wouldn't want any other, something she remembered all of her life.

On May 3, 1935, Ollie fell ill and was sent to the Army Navy General Hospital in Hot Springs, Arkansas. Details are not provided in either the official records or in Mildred's memoirs but he did not return to Wentworth duty until August 10, 1935. During this time, the Reed family took their little red trailerhouse for a month-long vacation traveling across the American West.[112] Just prior to his return, Ollie was promoted to the rank of major.

Buddy in his Wentworth football photo.

The illness did not deter Major Reed in the least. Alongside his roles as professor, trainer, and military leader at Wentworth, Ollie also tackled extracurricular tasks such as inaugurating and directing production of the academy's first Military Ball. On February 29, 1936, six hundred guests, including Missouri Governor Guy B. Park, attended the gala. "Honorary sponsors," young ladies recruited from various schools to attend the dance, participated in a photo competition judged by actor Gary Cooper.[113] Early Coleman and his eleven-piece band from Kansas City's Ambassador Hotel provided the night's music. The evening concluded with the singing of the Wentworth Song:

> Wentworth, Wentworth, Bless Your Heart,
> Wentworth That We Love So Well,
> We'll Always Be True And We'll Stand By You,
> Wentworth That We Love So Well.

At a time when the rest of the country was bending toward isolationism, the purpose of a Military Science program was to teach preparedness for combat, whether at home or abroad. The relatively idyllic life of Lexington, in spite of the deprivations wrought by the Great Depression and dust clouds from the barren Great Plains during the "dirty thirties," stood in stark contrast to the news from across the Atlantic: "Once More, Europe is Ready for Big War." Ollie knew that America could again be drawn into conflict on foreign soil, and he had the responsibility of training young men and their leaders to survive on the battlefield.

Wentworth underwent an annual inspection by the War Department, represented by the Seventh Corps Area overseeing forty-three ROTC programs in eight states, for accreditation as an Honor Rated School. Only schools with Regular Army officers detailed as professors of Military Science and Tactics could qualify as honor schools. Then, from each of the honor schools, students could be designated as honor students by the professor of Military Science based upon their academic record and leadership qualities. This appellation would support their admission to a service academy, such as West Point, or in their commission to the Regular Army.

Wentworth passed this test every year during Reed's seven-year tenure. Inspections also included one in 1936 by Secretary of Defense Harry Woodring, a former governor of Kansas, and the 1937 inspection was conducted by Major James E. Jeffres and Major William G. Livesay. Jeffres represented the Atchison, Kansas congressional district for two terms in 1979-83, and General William Livesay would assume command of the 91st Division from General Charles H. Gerhardt in 1944.

Outside of Wentworth, Ollie led a busy life in the community. He was a member of the Knights of Pythias Lodge 197 and the Lexington Masonic Lodge 149. In Lexington, Colonel Sellers was a high-ranking Mason, as were many other local businessmen, and membership for Ollie meant good standing in both Wentworth and the surrounding community. In April 1935, one hundred Masons from the Masters and Wardens Club, and the Orient Lodge of Kansas City, traveled to Lexington to participate in the conferring of the "Third Degree" on then-Captain Reed. He attained the title of Master of the Third Veil the following year. He also was active in scouting, as well as the local Rotary Club, and the Lexington Presbyterian Church.

In August 1935, the *Lexington Intelligencer* noting Ollie's promotion to the rank of major, recognized his civic involvement in a front-page story: "Major Reed has taken an active part in community affairs as well as becoming a recognized leader of the activities of the young men of the community."

Mildred was an active member of the Lexington Women's Club. Not long after the Reeds arrived in Lexington, Mildred attended a Women's Club meeting featuring Mrs. DeWitt Chastain, state president of the Federation of Women's Clubs, speaking on the topic of "Men's Ideal of War Must be Destroyed." The next year, the club met on the topic of "The Best Cure for War is Elimination of Glory and Profit." In 1935, there was a change in the club's direction when Mildred was elected president of the Lexington Women's Club. The topic for discussion at her first meeting was flowers. As an officer's wife, Mildred shared her husband's apolitical nature. Officers did not involve themselves in politics, going so far as to not vote in elections, and the same, preferably, held true with their spouses. Anyway, a women's club, to Mildred, was a place for lighter, more feminine discussions, rather than politics.

She also enjoyed speaking on religious themes. During one such presentation, "The Living Christmas Tree," she decorated a Christmas tree while telling the significance of each Christian symbol among the ornaments. Mildred was not preachy or dogmatic about religion. Rather, she was a very tolerant and understanding person who was quietly confident in her own faith and tolerant of others. Never in her memoirs or letters did she express an opinion about anyone's religious beliefs, including her own, other than writing of her deep faith in God.

Meanwhile, eleven-year-old Ted was an avid member of the Boy Scouts. Ted and other boys from the Lexington troop were constant features in the local newspaper for their accomplishments. Ted was the leader of the Skull patrol in

Lexington's Troop 318. Membership in scouting was high during this period, but few of the boys made it as far as Ted to the rank of Eagle Scout.

Ted, right, and two friends are packed for a camping trip, leaving on foot.

Lifelong Lexington resident John Morrison remembered Ted in 1939 as a scout leader who took several scouts on a fourteen-mile hike, a requirement for the first class scout badge. "Ted's dad drove a carload of scouts to Higginsville and Ted hiked us back to Lexington on the old Higginsville road. As I recall we stopped at several farmhouses along the way for a cool drink. Most memorable was waving at a train and the train stopping to see if they could help in any way. Ted was a great leader and all made it back home without too many blisters."[114]

Because of the Depression, many of the boys advancing through their teen years had to work in order to help their families, leaving little time for Boy Scouts.

Left, Ted with his guinea pigs; center, studying chicks; and with his puppy in Lexington, Missouri.

Ted's love for animals continued to grow from his horseback riding at Fort Benning. Mildred recalled her son's fondness for animals and how people in Lexington always thought of Ted and his guinea pigs, Hiram and Minnie, who furnished offspring for half the youngsters in Ted's age group.

Ted attended the Arnold School, one of Lexington's four public schools. It was the elementary school for white children. Education in Lexington, as it was statewide, was "separate but equal." This was legislated across Missouri after the Civil War and upheld by the state Supreme Court in 1890.

Arnold School, 1933. Ted is on the far left, top row.

The unanimous decision against the African-American students was penned by Justice Francis M. Black, noting that "color carries with it natural race peculiarities," and that these differences could never be eliminated, justifying separation of blacks and whites.[115]

It took a 1938 U.S. Supreme Court decision to overturn the University of Missouri Board of Curators denial of African-Americans the right to attend state universities.[116]

When the Reed family moved to Lexington, Ted was extremely shy and reticent in his new surroundings. He tended to stand apart from the others and wasn't quick to make friends. He had left his good friends from two years at Fort Leavenworth and now he had to start all over in a sixth grade class full of strangers. For Buddy, the task of making friends was made easier by boarding at Wentworth, but Ted was on his own – typical of military life. By eighth grade Ted was popular enough to be elected class president. Scouting truly invigorated Ted's interests and widened his circle of friends. His father also was very active in scouts, serving on the local scouting committee, and as a merit badge counselor in marksmanship and bugling.

A highlight for Lexington scouts was a speech delivered by cowboy movie star, Tom Mix, at a rally in Lexington's Central College Park. Mix brought his Circus and Wild West Show to Lexington, speaking to the boys on the importance of scout craft. It was after this speech when Mix rode his horse Tony into Pat's Bar on South Ninth Street.

In the summer of 1935, eighteen-year-old Betty Sarvis came to live with the Reeds. Betty was the daughter of Joe and Christine Sarvis. It was Christine who helped Mildred get settled in Manchester,

Ollie, Buddy, Mildred, Ted, and Betty in Lexington, MO.

Connecticut when Mildred was pregnant with Buddy, dying soon after returning to Norton. And, it was one-month-old Betty, as Mildred remembered, who was the only one to cry at her wedding. In 1919, by the time Betty was two, her mother passed away. Later, Joe married Ruby Wright, and they lived in a small bungalow at 731 Haines Street in Dallas with Betty and her older sister Louise. A younger half-brother, Stephen, was born in 1924. Throughout her teenage years, Betty constantly fought with her stepmother. The two did not like one another and pulled no punches at home. Ruby died of pneumonia in April 1934 and Joe quickly realized that he was overwhelmed with three children, two of them teenage girls. When Mildred asked if there was anything she could do to help, Joe sent Betty to Lexington immediately after her 1935 graduation from Oak Cliff High School in Dallas.

Betty was a wild child, but she and Buddy were kindred spirits. Normally, he was rather shy but Betty had a way of teasing the wilder side out of him. Betty was a very outgoing person who acted in high school plays and sang in the chorus. Bud and Ted met Betty when they visited Texas in the summer of 1931, and the three became quite close. In fact, the two boys had crushes on their older cousin. Four years later, Buddy and Betty, almost two years his senior, became "kissin' cousins" as they spent as much time together as possible when Buddy was home from

Buddy and Betty in Texas, 1931.

school. The Texas girl in ankle socks and a pretty bow in her hair had grown to become an attractive young woman with tight curls hanging over her forehead and a slightly gap-toothed smile. Betty's angular facial features enhanced her thin frame to almost delicate. Four years earlier, Betty was slightly taller than Buddy. Now, he had grown taller than she. In one of their forays around Lexington, Betty flew out of a rumble seat in the car in which she and Buddy were riding, breaking her tailbone. Betty always remembered Buddy coming to her rescue. She was hopelessly in love with the handsome military school cadet.

Mildred put forth her best effort at refining Betty, and the lessons she taught Betty were treasured by her student forever. Posture, speech – every refinement. Betty was the daughter Mildred had always wanted, and Mildred was the mother Betty needed. To Betty, Mildred was Mommie-Mit for the rest of her days.

In the summer of 1936, Betty won a contest while working at the Connor-Wagoner store in Lexington. Fifteen employees in the three women's wear stores competed for the prize - a weeklong, all-expenses-paid trip to Dallas and the 1936 Texas Centennial aboard an airplane. Each contestant was awarded one point for every dollar in sales, and Betty was the Lexington winner. She and the two other winners had never flown before boarding a Braniff Airways plane, nicknamed the "Centennial Flyer," on Sunday, August 9th at the Kansas City Municipal Airport for the four-hour flight to Dallas.

"I can't hardly believe that I really won," Juanita Ward, a clerk at the Warrensburg, Missouri Connor-Wagoner store, told the Warrensburg Star-Journal. "Now maybe I can get some sleep for a couple of nights."

Within a year, Betty was living in Los Angeles and, on June 26, 1937, she married Charles Lorton, a sales clerk at the Barker Brothers Furniture Store in the city.

At the same time Betty was living with the Reed family, Ollie's mother had arrived from Norton in October 1935. She fell ill the following January and three months later, on March 23, 1936, seventy-year-old Mary Plusky Reed died in Ollie's home at 1722 Main Street. She passed away from chronic myocarditis, an inflammation weakening the heart muscle that could have developed or worsened during the dust storms sweeping the Great Plains. Her husband, Orville, had passed away in 1921. Ollie took his mother home to Norton for burial in the town cemetery.

From Main Street, the Reeds moved to 1909 South Street, three blocks from the Wentworth campus. A block away from the Reed home, a neighborhood girl caught Buddy's eye at 1824 South Street. Nancy Campbell lived with her grandmother, Mattie Yingling, a twice-married boardinghouse proprietress who inherited the South Street property from her late, second husband, C.E. Yingling, a tinner by trade and longtime member of the Lexington school board. Mattie had two children from her first marriage to Dan Frazier, Lola and John.

Lola Frazier and Captain Ralph Campbell
(Photos courtesy of Paula Frazier)

Lola Frazier married Ralph W. Campbell in 1914. In the mobilization for the World War, Ralph organized Company A of the Sixth Missouri infantry. He was promoted to the rank of major and given command of the First Battalion of the 138th Infantry Regiment, 35th Division. The 138th Regiment of the Missouri National Guard was involved in some of the heaviest fighting of the war in the Alsace, Loraine, and Meuse-Argonne campaigns, in which Campbell was cited for gallantry. In the battle of Meuse-Argonne, Campbell fell under a cloud of mustard gas.

After the war, Ralph suffered mightily from the effects of the gas. Breathing was difficult and he was often ill. He tried his hand in the hardware business buying Samuel Drysdale's store at the corner of Tenth and Main in August 1919 with Raymond Mayer, but the business failed. Ralph next worked at the Wentworth Academy and then later joined the Lexington Police Department.

On July 26, 1921, Lola gave birth to Nancy Elizabeth Campbell. Nancy's parents loved her dearly. Lola, a thin woman with soft and gentle features, was a Lexington public school teacher. Two days before Christmas in 1932, Lola fell ill with pneumonia. Five days later, at four o'clock in the morning, she died in her bed at the age of forty-one.

Love and Sacrifice

Ralph was devastated. He fell into a deep depression, made worse by heavy drinking. He and Nancy moved out of their house on Main Street and into his mother-in-law's South Street home. Three years later, Ralph Campbell, forty five, died of cirrhosis of the liver on the day after he checked himself into the Veterans Hospital in Excelsior Springs. The local newspaper ran a front-page story about the passing of a local war hero. Nancy, now fourteen years old, was suddenly an orphaned only child. She remained in her grandmother's house. Mattie Yingling was a stern guardian whose husband, Charles, had died only one month before Nancy's father.

As Buddy was two years older than Nancy, he was slow to recognize the girl down the street. He was often busy with Betty Sarvis when not living at school and Nancy was quite shy, as well. She had inherited her thin figure and calm demeanor from her mother. In her junior year, Nancy was noted in her high school yearbook for her coveted hairstyle, a chic brush curl. Her poise and long neck also set her apart from the other girls. She was tall, though not the tallest, but certainly a head taller than Bud who stood at barely five foot five.

In 1936, Nancy was a member of the first Lexington High School Pep Squad that debuted with a pre-game parade through nearby Higginsville where the Lexington Minutemen would be playing the Higginsville Blue Jays. Resplendent in their blue corduroy culottes, white shirts, and red jackets, representing the school colors, the squad, "marched, paraded, formed circles and cut fancy figures in their flashy new uniforms," as described by the school newspaper.

1938 Lexington High School Pep Squad. Nancy Campbell is at the center of the middle row.

Nancy was very quiet and reserved, but throughout high school and college, she was a champion debater. Debate in the 1930s was just as popular as any sport, and this was especially true in and around Lexington. The local team regularly appeared before groups of farmers and local fraternal organizations to argue given pros and cons.

On Saturday, March 5, 1938, Nancy and her "affirmative" debate teammate, Marjorie Sue Bell, brought home one of the greatest triumphs in Lexington educational history by winning the state debate tournament, continuing on to represent Missouri in the national championship.

An "Incubator" column in the Lexington High School newspaper, written by classmate George Canning, praised the pair: "I think that every Senior should be feeling very proud these days. Think of being able to say, 'I'm in the class that has those two swell debaters in it'. Of course you know I'm talking about Nancy Campbell and Marjorie Sue Bell."

In Leslie Bell's 1962 history of Lexington education, the high school debate team's accomplishment is one of only two student achievements noted in the school system's history from 1853 to 1959:

Lexington High School debaters placed first in the State National Forensic Tournament at Liberty on March 3 and 4, 1938, qualifying them for the national tournament in Wooster, Ohio. The affirmative team was composed of Marjorie Sue Bell and Nancy Campbell; the negative Audrey Adams, Edwin Ragland and Marilyn Hicklin. Joe C. Amery was the coach. The Lexington team placed 18th in the field of seventy at the national championship, a commendable showing.[117]

After Betty Sarvis moved to California, Buddy and Nancy grew close. How they met, exactly, is unknown and the two shy teens did not make a show of their relationship. A benefit for Nancy was that Mildred adopted her as a daughter. Mildred loved her two sons but longed to have a daughter she could help mold, just as her mother shaped her. In Betty, she took a wounded young lady who had drifted from the straight and narrow, and helped her find her way. With Nancy, Mildred, once again, provided maternal love for a girl who lacked parental love and attention during her formative years. Mildred provided a comforting beacon for Nancy to follow, reinforcing lessons being taught by her grandmother, but with a more gentle hand. Buddy and Nancy's relationship accelerated as the time drew near for Bud's departure from Wentworth. Buddy was graduating from high school in 1936 and had received acceptance to take the entrance exams for West Point. Nancy would graduate two years later but she knew her future was by his side.

So did Mildred. "Buddy-and-Nancy were one word during high school," she later recalled.

The summer of 1936 came and went. Buddy graduated from the Wentworth high school and would be returning to Wentworth for at least another year in the junior college program. He had traveled to Kansas City for the West Point exams but failed to make the grade. This was not unusual and he could apply for entrance the following year.

★★★★

In 1937, economic conditions improved. The New Deal programs were working and President Roosevelt was sworn in for his second term, sweeping the 1936 election against Alf Landon, governor of neighboring Kansas. The spring wheat crop was the best in six years.

In September 1937, the recovery in Lexington was set back when a cornerstone of the downtown area burned to the ground. The Traders Bank building, a Romanesque Revival structure built in 1892, that had been transformed into The Great Atlantic & Pacific Tea Company (A&P) grocery store, was destroyed, symbolically marking the end of an era in Lexington. Police had to erect barricades around the grocery building with no trespassing signs to keep people from what they called "salvaging" goods.

Beyond the streets of Lexington, international aggressions were spreading. The United States remained neutral, maintaining a "cash-and-carry" policy for European nations, particularly weapons and machinery sold to England and France, under a series of Neutrality Acts signed by President Roosevelt. The 1937 act, signed into law only days after the German Luftwaffe destroyed the Spanish town of Guernica, specifically kept American weapons and personnel out of the Spanish Civil War. Americans, meanwhile, invaded Europe as an army of tourists.

Asia was a different story. On July 7, 1937, Japanese and Chinese troops exchanged fire, opening the Second Sino-Japanese War. Japan soon overran and occupied Peking and Tientsin. Atrocities were widespread on both sides of the conflict that had flared up from time to time since the nineteenth century, but 1937 marked the beginning of aggression that would not end until the cessation of the next world war.

Bud completed his first year of the Wentworth junior college, took the West Point exam again, finally receiving the letter he, and his father, had long awaited with word of his acceptance to the academy. The letter arrived the same day as the Wentworth commencement.

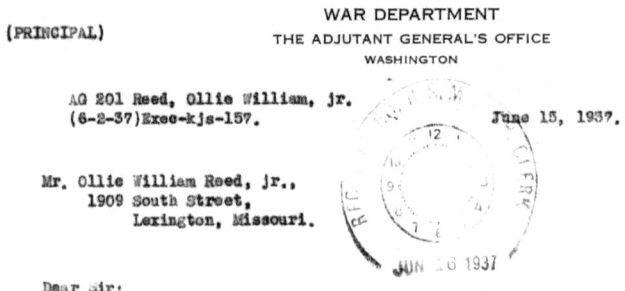

Ollie and Mildred were proud enough to burst when word arrived. Neither wanted to impose their personal desires for Bud's future upon him, but Ollie was aware of the benefits his appointment would have on his son's future. Mildred kept her expectations in check as a firm believer the events occur in a natural course under God's watchful eye. But, Bud had grown up along the lines of a private rising through the ranks and this was a significant promotion. He loved his father, the major, with all his heart and had great respect for him, just as a young lieutenant has for a higher-ranking officer who treats his men as his very own sons. That was Ollie's style of command – easy going, and firm but fair. As he was away so often, the real responsibility fell to Mildred to raise the boys, and it was certainly the hardest on her to say goodbye to her eldest son.

Bud had less than two weeks to pack and report to West Point on July 1. He was gratified with the acceptance, but just like his father in 1915, he was also deeply conflicted about leaving his family and, especially, Nancy. His parents knew their son was torn, but they would not force him to go, though his father made impassioned and persuasive arguments on behalf of the academy. Ollie may not have voiced personal opinions about politics but he knew what he wanted for his son. Knowing his responsibility and duty, Bud joined his father in packing the car for the twelve-hundred-mile drive to Highland Falls, New York and West Point. He knew how much his acceptance meant to his father, so in spite of trepidations over leaving and entering another regimented environment, Bud kept his misgivings to himself.

Everyone missed Bud terribly, but life had to go on. Nancy traveled to Kansas City, where she spent the summer with her Uncle Cyrus Campbell and his family. She still had another year of high school, graduating in 1938 from Lexington High School. She left Lexington to attend William Jewell College in nearby Liberty, Missouri. Nancy accelerated her schedule to be in sync with Bud's scheduled 1941 graduation from West Point.

Nancy Campbell graduation, 1938, Lexington High School
(Courtesy Paula Frazier)

Over the summer of 1937, Mildred taught at the Lexington Presbyterian Church vacation bible camp and Ted spent several weeks at the Osceola Boy Scout camp. After taking Bud to West Point, Ollie traveled to Fort Leavenworth with Colonel Sellers to inspect the Wentworth cadets at the summer ROTC camp, and then Ollie continued on to Fort Riley, where he participated in army maneuvers. It was a relief to be back in the regular army, if only for six weeks.

When he returned, Ollie gave a speech to the Lexington Rotary Club about the maneuvers and the preparedness of America's military. In his address, he paid particular tribute to the Air Corps. He also issued one portentous caution: "There is one lesson," he said, "that always has to be learned – when communications break down, the attack breaks down."

In April 1939, Major Reed received orders to sail to the Philippines on September 12. He had managed to stay on at Wentworth long enough for Bud to complete high school and gain admission to West Point as an Honor Graduate of Wentworth. Ted had only one year remaining of high school, which he could complete overseas.

After he school year ended, Ollie stayed in Lexington to assist with the ROTC camp, sponsored by the army's Seventh Area Corps, and to help in the transition for Major R. J. Williamson, who would succeed Ollie at Wentworth.

Throughout May and June, events honoring the service of Major and Mrs. Reed to the Lexington community were held. Fifty people attended a dinner in Major Reed's honor, sponsored by the Lexington Turner Society, where Ted led the Pledge of Allegiance and demonstrated wood crafting. On June 1, the local Masons held a dinner where Ollie was presented a "handsome Masonic watch charm."

In late June 1939, sixteen-year-old Ted attended Camp Osceola Boy Scout camp (now the H. Roe Bartle Scout Reservation) in Iconium, Missouri for the final time. At the close of the camp, the scouts sang the new camp song, "The Hills of Osceola," written that year by Edmund Wilkes Jr. Its lyrics can still be heard today being sung by Missouri scouts in their Ozark campground:

> The hills of Osceola are calling me today,
> Come back along the Scouting Trail, their voices seem to say.
> I dream of woodland valleys, and pathways that I know,
> And answer, O-sce-o-la dear, I'm coming back to you.[118]

The day before his family's departure from Lexington, Ted celebrated his seventeenth birthday with his closest friends. Norman White,

Bob "Kingfish" Morrison, and Ned "Dude" Barnet were joined by Shirley White, Nancy "Bitty" Aull, and Ted's girlfriend, Virginia Lee Beissenherz. Ted had shaken his shyness, becoming class president, running track and playing on the Lexington High School Minuteman football team. The good friends enjoyed lunch together for the last time, having their pictures taken in various groups before finishing up at Freddie Anton's family ice cream shop.

Virginia Lee Beissenherz, Ted, Shirley Russell, Ned Barnett, and Bitty Aull at Ted's 1939 birthday.

The Japanese were engendering hostilities with the Soviet Union and China, while Hitler annexed Czechoslovakia and the Sudetenland into his Reich. President Roosevelt was calling for the mobilization and rearmament of America. It was time to return to the army.

On July 27, 1939, a front page story on the *Lexington Intelligencer* told of the departure of Major and Mrs. O.W. Reed, and their son Ted for West Point, where they would visit Buddy before sailing for the Philippines. While Major Reed's accomplishments at Wentworth were highlighted, Mildred's civic activities drew a special note:

> The departure of Major and Mrs. Reed will be greatly regretted by their many friends here. Both have been prominent in activities aside from connections with the school. Mrs. Reed has been a leader in church, club, and social life and in a variety of civic enterprises where her ability and willingness to serve have been of much value to the groups with which she has been connected.

Buddy beside the Missouri River, Summer 1937.

THE BURLINGTON HOTEL
380 ROOMS FIREPROOF
VERMONT AVENUE AT THOMAS CIRCLE
WASHINGTON, D. C.

Tues. night.

Mother Dear;

Write me a letter, please. To Cadet Reed — Pleble Camp — U.S. Military Academy — West Point — N.Y. — Ain't that grand. I'm so happy, I could walk on air. And Mother more than ever before, I want to congratulate you. You married the grandest man in the world! If you could have only seen him these last 9 days. Walking, walking from desk to desk, office to office; talking, talking — every time with men who rank him! He

CHAPTER ELEVEN

Duty, Honor, Country - West Point

USMA Cadet Ollie W. Reed, Co. G,
October 1937

OLLIE AND BUD packed the car for the drive east to Highland Falls, New York and West Point, with a stop in Washington, DC where Major Reed would lobby army brass on behalf of his cadet-candidate. A physical exam at Fort Leavenworth threw a roadblock into his path – Bud was found to be colorblind. Ollie was not about to let this test deny his son the opportunity of a lifetime.

Arriving in Washington on Sunday, June 21, 1937, Ollie set to work early the next morning. Walking the halls of the State, War, and Navy Building,[119] located next to the White House, Major Reed started at the top, demanding to see the Army Chief of Staff, General Malin Craig, West Point class of 1898, and the adjutant general, Major General Edgar T. Conley, a 1897 graduate of West Point whose son also graduated from the academy. Ollie was dogged in his determination. After some goading, General Conley finally acceded to forward the case to the Surgeon General with the general's resolution.

"Tell the Surgeon General's office to give this boy the yarn test and abide by it."

The yarn test is one of three major color perception tests that groups similarly shaded strands of different colored yarns, arranged by color group, hanging in equal lengths from a stick. The subject is handed a piece of yarn and asked to match it to the strand on the stick. Bud passed with flying colors, so to speak, only two days before he was to report to West Point.

Bud excitedly wrote home to his mother on his last night at the Burlington Hotel in the District.

> If you could have only seen him these last 9 days. Walking, walking from desk to desk, office to office; talking, talking – every time with men who rank him! He bullied those Colonels and Generals around as if they belonged in his company. He practically forced them to give me an O.K. He did for me something he won't do for himself. He could easily talk himself into War College (and he does want to go!) if he worked that hard for himself. And I can't thank him! There's nothing I can say to thank him! What can I do? I helped him as much as I could, but half the time I was running to keep up with him. And you could just see, day by day, their attitude changing. He got everybody in Washington wanting me to pass; and as I took my final test, all the doctors and nurses were there rooting for me. It's simply marvelous. I only hope I'm half the man my Dad is! And someday I'll try to do for my kid what he has done for me. That's the only way I can thank him. That and pass the Point with honor!
>
> Incidentally, I guess the grandest man in the world got the best girl in the world for his wife.

After stopping in Merchantville, New Jersey, to visit the "tucco" house ("It is still home," Ollie wrote to Mildred), and old friends Captain Andrew "Mac" McCully and his wife, Catherine, they spent the night in Morristown before driving the final sixty five miles to Highland Falls, New York, arriving at West Point in the morning of July 1, 1937.

"At 9:01 AM an M.P. told Ollie, 'This is as far as <u>YOU</u> can go,'" as wrote to Mildred later that day. "Bud & I shook hands said Good-bye and I haven't seen him since. It was the most sudden and complete severance of ties that you can imagine."

Ollie was invited to lunch at the West Point Army Mess, where he

Love and Sacrifice

reunited with First Lieutenant Ralph Woods, who was fresh out of the academy and the class of 1929 when he joined the 29th Regiment at Fort Benning when Major Reed was commanding Company H.

At West Point, Woods was a drawing instructor who later would be assigned to the same regiment as Bud. They were joined by another young officer, First Lieutenant Thomas Wells, son of General Briant Wells, superintendent of the Infantry School during Ollie's stint there. Wells was an assistant instructor in tactics and a company commander of cadets, a "tac" who would keep an eye on young Cadet Reed.

Ollie decided to stick around until the parade and the swearing-in of the "new boys" afterward.

"Hope I can recognize Bud but will feel better at any rate…I'm lost – but guess I'll recover," he closed his letter to Mildred. "Of one thing I am now satisfied, 'Bud wanted to come.'"

His father may have been able to lobby for his admission, but now, inside the gray walls of West Point, Bud was on his own.

Bud was directed to walk through the campus to Thayer Hall. This was the beginning of Cadet Basic Training, or Beast Barracks, the six-week indoctrination of cadet candidates hoping to gain the honorary title of Plebe, from their lowly initial anointment as an Animal or Beast. A rigorous introduction to academy life, the physically and mentally challenging period was divided into two portions – the first three weeks were on campus learning the basics of discipline and military drills, followed by three additional weeks living in tents on the edge of campus overlooking the Hudson River where Plebes would practice marksmanship and other military basics. Throughout, they lived a Spartan existence while their second-year bosses, known as Yearlings, enjoyed visits from girlfriends (known as "femmes") and limited freedoms. Academic classes began at the conclusion of Beast Barracks.

Cadet candidates started arriving at 8:30 a.m., first reporting to the entrance to the Gothic-styled Administration Building (today, Taylor Hall) where their credentials were examined. Next, they were directed across a campus street to the West Academic Building where they surrendered all of their money and any "contraband," including such items as alcohol, playing cards, and weapons. In the same building, they were directed to a side room with benches, adjacent to another room filled with large examination lamps. Told to strip down to their underpants, the cadet candidates were given a brief physical exam for signs of infectious or contagious diseases, dressed, and sent outside to the Area, a central courtyard surrounded by the gray stone buildings of the Central Barracks. There, cadet candidates were ordered to their company area and immediately surrounded by a small, barking army of second-year cadets dressed in gray and white.[120]

R-Day 1937 (Courtesy of Anne Allen)

The Beasts were mockingly addressed in shouts. "Shoulders back! Chest up! Get some more wrinkles in there, Mister Dumbjohn. Stomach in!" a Yearling screamed into a Plebe's face as he was surrounded by three other barking upperclassmen.

This was R-Day, or Reception Day, a time when new cadets would receive a one-day crash course in Beast Barracks.

"Eyes to the front, Mr. Ducrot! Get in step, Mr. Dumbguard! Don't fall down in ranks, Mr. Duficket! Bounce that chest up, Mr. Fluzfoot!" Yearlings ordered each of their charges to remove their jackets and ties, and sometimes had them roll up their pants to their knees.

The young men were put into groups according to height. As many as five Yearlings surrounded one Beast, ordering him to pick up his suitcase, put it down, pick it up again, and so on. Repeating this and other motions, Mr. Phlogg would otherwise have to stand at attention in a position known as "finning out" – shoulders back in the "position of a soldier" with his arms clamped close to his body and his chin driven into his raised chest. During hazing in years past this posture was known as "bracing;" a Plebe would throw his shoulders back until the blades met, draw in his chin, suck in his stomach, and walk so his toes touched the ground before his heels for long periods of time.

Love and Sacrifice

The goals of this training were twofold: one, that rank has its privileges, and two, that no matter how impossible the order, it could be accomplished.[121] It also served to strip away any vestige of ego and self-importance. The intended takeaway: an upperclassman never thinks of his rank and a subordinate never forgets it.

Cadet candidates were constantly reminded of the simple path to success: "Keep your mouth shut and your ears open," and their station in life: "You're in the Army now!"

Likewise, from the very beginning, the words of "Duty, Honor, Country" were instilled in the cadets as the guiding principles they would carry through the academy and life.

As they were grouped by height, the tallest were in A and B companies, while the likes of Bud, standing a little over five feet five inches (the minimum height requirement was five-feet four inches), were in Company G, a "runt" company. After that came late arrivals, returnees, athletes, and others, all the way to Company M. Cadets spent practically every waking hour over the next four years with members of their company and little time with any others.

Cadet Corporal Jay Beiser marches newcomers on R-Day 1937 (Courtesy of Anne Allen)

"Don't say Companny [sic] – say 'Ko.' 'G' Co. men or 'H' Co. men," Bud later wrote home to his mother. "We in 'G' Co. are always called gnomies (pronounced Guh'-nomies by them) by 'flankers' and despised as such. The feeling is reciprocated. There is also a lot of rivalry between the Cos. And Battalions. There are 3 battalions 1(A,B,C,D) 2(E,F,G,H) and 3(I,K,L,M) No 'J' Co. as usual in the Army."[122]

Each group reported to the cadet first sergeant and then double-timed to the Cadet Store where they picked up white shirts and gray pants with the distinctive black stripe down the side, as well as their full-dress grays and new "skins" – gray trousers and gray shirt, a feather-duster hat (known as a "tar buckets"), and a fifteen-pound full-dress overcoat with forty-four brass buttons, with a wool cape draping the cadet's shoulders.

The young men immediately learned the West Point maxim that "cadets don't walk, they run," with each Beast running as fast as he could, laden with all of his newly-issued clothing to his assigned barracks, quickly hanging each item in place, ready for inspection. As soon as the last item was in its proper place, the Beast ran back to the Area, where he was handed another load, this time bedding, a rifle, and packs loaded with field gear that had to be rolled up in the mattress and then lugged over his shoulder, as fast as possible, to the barracks for unpacking. Without hesitation, it was back to the Area, dressed in new gray shirt, gray pants and white waistband, ready for the five-thirty march to the Battle Monument at Trophy Point overlooking the Hudson River. There, they were sworn in as official cadet candidates by Captain E.L. Sibert of the Field Artillery,[123] who read the oath, which was repeated en masse, word for word, each inserting their name:

> I, Ollie William Reed, Junior, do solemnly swear that I will support the Constitution of the United States, and bear true allegiance to the National Government; that I will maintain and defend the sovereignty of the United States, paramount to any and all allegiance, sovereignty, or fealty I may owe to any State or country whatsoever; and that I will at all times obey the legal orders of my superior officers, and the rules and articles governing the Armies of the United States.

The oath was signed, notarized, and placed in Cadet Candidate Reed's personnel file, where it remains on record today.

One of the most important items given to any Beast was his copy of *Bugle Notes*. This was, and remains, the Plebe bible, containing all the trivial information a new cadet must learn because, somewhere along his first year, an upperclassman will, no doubt, suddenly stop him to ask a random question out of the blue.

"How many lights in Cullum Hall, Mr. Ducrot?" (340)

"How many gallons in Lusk Reservoir, Mr. Dumbguard? (Seventy-eight million gallons when the water is flowing over the spillway.)

"How many names on Battle Monument, Mr. Dumbjohn?" (2,230)

Likewise, a Plebe could suddenly be asked, "What's your P.C.S.?" meaning, did he work before coming to the academy, or "previous condition of servitude." If the Plebe had not worked a job beforehand, his response would have followed a line such as: "My previous condition of servitude was school girl, sir."[124]

Nothing was sacred from the upperclassmen, especially not candy sent to cadets. Bud wrote home about the thievery of mailed treats, telling his parents that perhaps they shouldn't send any food in the mail: "They'd let me have one bite and then they'd say 'Are you sure you wouldn't like some more, Mr. Reed?' That's my cue to say, 'No Sir, help yourself!' If I don't, woe betide."

Mail with any bulk or substance was fair game for the upperclassmen but the Reeds put one over on them, as Bud wrote home: "There were 4 of the most disappointed upper classmen I ever saw when they opened my package and found only books!! Ha!! Ha!! I just couldn't keep from laughing. I had to stand against a wall for 15 minutes. I still laugh when I think of it."

Using the knowledge of the upperclassman's aversion to mailed books, Mildred devised a method of sneaking candy in a hollowed-out book.

"Thanks for the caramels," Bud wrote her. "The book worked swell. I'll send it back and we can try again. I saw another boy who got a package of candy in the same mail as I got my 'book'. He only got about 5 or 6 pieces of his 'boodle'. I guess we sneaked one past 'em. They just tore a corner of the paper off to see if it was a book. I almost wrecked the book getting at the candy."

Hazing was officially eliminated from academy life at the turn of the twentieth century, so the treatment Bud and the other Beasts experienced was tempered. Sanctioned hazing ended following the death of Oscar Booz, who entered West Point in 1898 but resigned after four months of torment. Booz died of tuberculosis eighteen months later, and his family's outrage sparked a nationwide scandal, accusing West Point and its hazing ritual of causing their son's death.

Around the same time as Booz's death, Cadet Douglas MacArthur was mercilessly hazed as the son of Brigadier General Arthur MacArthur and for being a "mama's boy." In spite of this, MacArthur agreed that the goal of hazing was worthy but that the method in which it was employed was too often violent and uncontrolled.

These introductions to West Point life earned the academy such monikers as "Hell on the Hudson" and "Alcatraz on the Hudson." The pressure was unrelenting, and many cadet candidates did not survive the first three weeks. Army brats, like Bud, and those with military prep school training had an easier time adjusting to the physical and mental stress, as they were already somewhat familiar with military discipline and the rigors of training. It may not have been their first introduction to military life, but it was a very tough second.

Formation in the Yard. (Courtesy USMA)

The day began at 6:20 a.m. with the blast of the Hellcats, an ensemble of buglers and drummers, loudly rousing the Plebes into formation by 6:25 in their gray shirt and pants. After inspection, the cadets were dismissed with twenty minutes to clean up their rooms before morning inspection, followed by a quick breakfast and then five periods, about an hour each, of close-order drills, guard duty, physical drilling, and manual weapons. Finally, there was a break for dinner, and then twenty minutes for shining shoes and, if the time could be spared, writing a letter home. By the time Beast Barracks ended, the academic calendar was a relief.

During Beast Barracks camp, Bud was already writing to his father about how some of the training was not up to standards:

> I really don't think there's enough instruction. They explain a new movement and then practice it with no more comments than 'Let's go, now!' Another thing I can't understand is that the way a man is 'made' depends on the number of reports he turns in. In our Company [sic] they've 2 men who want to be corporals,

Love and Sacrifice

therefore they 'skin'[125] about 10 men every meal formation whether they need it or not. Incidentally, I'll be on the Area next month![126] I've only got one demerit leeway to last until plebe hike starts, 8 days from now.

"Who are you and where do you come from?" would often lead a Plebe to speak with pride about his military pedigree. From then on, the Plebe would be called "Army boy" and that was a sure way to be placed under the care of the son of an NCO, who would get any notions of self-importance out of the Plebe's system in a hurry.

Some cadet candidates unsuccessfully tried bucking the system. When asked, "How soon are you going to change," a flippant cadet might answer, "Sir, I always hope to be the man that I am now," when he should have simply answered, "Immediately!" Worse was speaking out of turn, as Plebes could only speak when spoken to. Those who didn't were simply called "B.J." - they would wash out "before June."

Infractions were enforced by cadet officers who issued "slugs" - demerits, walking tours, and confinement to quarters as punishment. Demerits impacted a cadet's class ranking, upon which everything depended. Each walking tour meant an hour marching alone in the Area with a rifle on his shoulder, back and forth, with nothing to stare at but the surrounding gray stone walls. Each confinement meant one hour in the cadet's quarters, whether a tent or barracks.

Cadets who had been big men on campus at high schools or other colleges had the most difficult time adjusting to the degradations of the lowly Plebe life. An officer's son could easily fall into the trap of vanity. Now, they were the smallest fish in a pond full of sharks. Cadet candidates were learning a fundamental dictum of army life from the very beginning: "Obey first, question afterwards," a lesson Bud and other officers' sons already knew and needed to remember.

With the start of the second phase of Beast Barracks, the cadet candidates packed their rucksacks for the short march to the Plain, a flat expanse overlooking the Hudson River, where they pitched a sea of tents. There, they learned basic military skills –weapons, marksmanship, navigation, rappelling, guard duty, tactics, the use of gas masks, all the while continuing their close-order drills, turning movements, squad formations, and manual arms exercises. Those who struggled with the training basics were assigned to the "awkward squads."[127]

Three hundred and twenty-four young men entered West Point through the Thayer Gate on July 1, 1937. Late arrivals plus "turnbacks," cadets dismissed the previous year for academic reasons and passed a re-entry exam, expanded the class of 1941 to 579. Of these, only 420 would graduate on June 11, 1941.[128]

As he left his oldest son on the heights above the Hudson River that day, Ollie hoped Bud would be among those graduates in four years' time.

During the American Revolution, the high vantage point over the Hudson River offered West Point, then known as Fort Constitution, a strategic advantage over British ships, which had to slow along the river's bends. General Benedict Arnold, who commanded the fort, had been a hero in the capture of Fort Ticonderoga in 1775, but by 1779, having been passed over for promotion, General Arnold was already scheming with the British. His plans came to fruition the next year when he was assigned command of the promontory post, which he would attempt to hand over to the enemy. When Arnold's plan was uncovered, he fled and joined the British, later commanding them against the Americans in Virginia. Upon the British surrender, Arnold fled to London, where he died in 1801.

President George Washington called for establishment of a military academy in his final Annual Message to Congress but it advanced very slowly through the one term of President John Adams, and then all hope was lost with the election of Thomas Jefferson, an outspoken opponent of a standing military.

Two months after his election, Jefferson surprised many when he established a United States military school at West Point, founded in "useful sciences," under the Military Peace Establishment Act. From its beginning, the foundation of the United States Military Academy at West Point has been as a school of engineering. Congress established the Corps of Artillerists and Engineers following the Revolutionary

War, and it was renamed the Corps of Engineers in 1802 when it was headquartered at the new military academy at West Point. Major Jonathan Williams, an engineer and grandnephew of Benjamin Franklin, was named the first superintendent. The United States Military Academy formally opened on July 4, 1802.

"Our guiding star," Superintendent Williams said, "is not a little mathematical School, but a great national establishment. We must always have it in view that our Officers are to be men of Science, and as such will by their acquirements be entitled to the notice of learned societies."[129]

As Stephen Ambrose wrote, "When Jefferson assumed the presidency in 1801, he was eager to found a national institution that would eliminate the classics, add the sciences, and produce graduates who would use their knowledge for the benefit of society. Within this framework Jefferson realized that a military academy had the best chance of success."[130]

Jefferson's desire to establish a school of higher learning was coupled with his drive to "republicanize" the officer corps that had been dominated by Washington's Federalists. Soon, this would be known as "Mr. Jefferson's Army." Upon its opening, the academy consisted of one teacher and twelve students, ranging in age from ten to thirty-four. Some were highly educated, others not so educated. Even after Williams became superintendent, textbooks were rudimentary, classes either went too fast or too slow, there were far too few instructors, and a library low on books as Congress refused to allocate funds for their purchase. Students and faculty often were selected on the basis of political affiliation, and not on their skills and aptitude. The curriculum adhered strictly to science, while philosophy and the classics were omitted. There was no limit to the length of time a cadet could remain at the academy.

Between 1802 and 1817, academics improved and West Point graduated 179 commissioned officers, many counting among the pioneers in American civil engineering, education, and commerce. They were responsible for the development of roads, canals and railroads crisscrossing the new country.[131]

Still, the academy languished as antipathy and outright antagonism toward a permanent academy endured. In 1810, enrollment stood at forty-seven cadets, three fewer than allotted by Congress. The approach of the War of 1812 changed the political mindset when it was realized that a new corps of military officers was needed. The cadet class was increased to 250 and the curriculum expanded to include military science, drawing, and French language courses. West Point was transformed into a true military academy. Once the United States entered the war, the number of enrolled cadets dropped drastically as they were called into service.

With the war's end, Captain Alden Partridge was appointed superintendent. His tumultuous two-year tenure came to a close when Sylvanus Thayer was appointed superintendent. President James Monroe and Secretary of War John C. Calhoun tasked Thayer with expanding the academy's mission. For his accomplishments, Thayer is regarded as the Father of the Military Academy, and his legacy is known as "The Spirit of West Point." One cadet wrote of Thayer: "His object was to make (cadets) gentlemen and soldiers."[132]

Many famous and infamous individuals have passed through Thayer Gate. Cadet Jefferson Davis, class of 1828 and later the president of the Confederacy, led fellow cadets in opposition to the closing of the annual Christmas party in what became known as the "Egg Nog Riot." In 1852, West Point graduate Robert E. Lee (1829) became superintendent. At the same time, James Abbot McNeill Whistler was a cadet before becoming a world-renowned artist. His father, George W. Whistler, was a graduate of the academy in the class of 1819, but Cadet James Whistler was expelled after his third year. Many Civil War leaders came through West Point, including Ulysses S. Grant, a member of the class of 1843; George G. Meade, class of 1835, commander of the Army of the Potomac; William Tecumseh Sherman, class of 1840, perhaps best remembered for his "March to the Sea"; Thomas J. "Stonewall" Jackson, class of 1846, a Confederate general; and George McClellan, also class of 1846, a Union Army general.

As the Civil War drew closer, sectional fights among cadets became more prevalent. Events escalated with the election of President Abraham Lincoln on November 6, 1860. Three days later, South Carolina called a Secession Convention, and, on November 19, 1860,

Love and Sacrifice

Cadet Henry S. Farley of South Carolina became the first cadet to resign over secession. Of 278 cadets at West Point on November 1, 1860, eighty-six came from southern states. Of those, sixty-five were discharged, dismissed, or resigned for reasons relating to secession.[133]

On January 23, 1861, Captain Pierre G.T. Beauregard of Louisiana was named superintendent of West Point only to be relieved of his post five days later. Less than three months after leaving West Point, Confederate Colonel Beauregard was in charge of the artillery in Charleston, South Carolina, that opened fire on Fort Sumter.

Following the Civil War, West Point began admitting African Americans. Henry O. Flipper was the first to graduate from West Point, in 1877.

★★★★

The second half of the six weeks of Beast Barracks was oriented toward combat skills, use of various weaponry, fitness, and managing conditions. After pitching their tents and unrolling their packs, the Plebes practiced rappelling, land navigation, including night marches, and the use of gas masks, under the ongoing supervision of the Yearlings. Calisthenics, running, and hill climbing were grueling tests of endurance.

The three weeks of Beast Barracks came to a conclusion with Camp Illumination when the sea of tents staked across the Plain were aglow in the nighttime darkness. The Plebes enjoyed a simple night of relaxation as the Yearlings celebrated with their "femmes" at a colorful costume party. For the Plebes, this night marked the end of summer camp and sleeping in the elements. The relative comfort of sleeping in the shelter of a dormitory awaited them.

Even in their spare time, Plebes had few privileges. They

Camp Illumination (Courtesy USMA)

had no leave and could not step outside the West Point campus. They were not permitted to attend any of the Saturday night "hops" (dances) that were held throughout the year. They could attend football games, including away games, but they had to earn that right. One such game was the 1937 matchup against Notre Dame at Yankee Stadium in New York City. More than sixty-five thousand people attended the game, including the corps of cadets, all of whom sat in a cold rain to watch Notre Dame win, seven-zero, thanks to a botched fake punt and reverse by Army. After the game, the fun ended for the Plebes. They boarded buses for the return to the academy while several hundred upperclassmen stayed in the city to attend the annual military ball, with co-eds from area campuses.

Bud wrote to his parents: "Everyone has from after the game until 12:00 off. All plebes 'fall out' and act like upperclassmen. The Boston trip is the best one. Everyone there likes the Corp immensely – much contrary to New Haven, New York & Philadelphia. Also it's an overnight trip. The Philly trip is much disliked as a trip. Of course the Army-Navy game makes up for it, but we have to leave there at 9:00 instead of 12:00."

Christmas was no exception as Plebes were confined to the campus, but that was the one time of the year when they could relax, free of the presence of upperclassmen. They hung out at the Boodler soda fountain, attended the Ice Carnival, and held a Plebe Smoker – a stag party without femmes. The one co-ed event was the New Year's Eve celebration they called, "A Long Corps for 1938!" Cadets had no ready cash, only a ten-dollar monthly "Boodle Book," which they could spend at the Boodler and in the Cadet Store.

Outside of these minor privileges, the life of a Plebe was unrelenting. Upperclassmen barked at them at every turn, and academic demands were stringent. Bud wrote home about the horrors of mealtime:

> Everytime a plebe asks (or rather shouts) for food, he gets his hands up. They throw bread, salt, pepper, sugar bowls, and any deep platter that food won't fall out of. They throw bread all over the mess hall – even trying to hit the O.C. sitting up in the 'poop deck.' Upperclassman talk at tables. They holler & yell & shout

& throw glasses & salt sellers & pieces of bread & anything else throwable. Between each bite you must drop your fork and sit at attention while you chew. When your mouth is cleared you may pick up your fork & take another bite. If you forget you 'sit up' for 5 minutes. And very few 5 minutes of 'sitting up' will leave you hungry at the end of a meal.

Mildred remembered one of the hard lessons she taught her sons that came in handy for Bud at West Point:

> What a Mean Mama, but I was rewarded for insisting that my children eat everything. When Bud went to West Point he wrote, "I'm glad you made me learn to eat everything. There are days, here, when we have cabbage, rutabega, squash - things that finicky cadets won't touch, so they go hungry."

Bud earned his fair share of harsh penalties for even minor infractions. By August 1937, barely a month after arriving, Bud had racked up thirty-six demerits for such things as a rust spot on his coat, a waist plate improperly shined, and on August 6 he was handed six demerits in short order for improperly shined shoes under his bed, an oily bolt, and rust on his rifle. By the end of his Plebe year, Bud was ranked 475th in his class for conduct, with 175 demerits. He left some shoe leather on the Area ground and a wore a deep groove in his shoulder from carrying his rifle on punishment tours, endlessly marching back and forth without purpose.

Academically, Bud quickly found himself in trouble and did not withhold the truth from his father.

> I feel that I ought to talk to you about my academic work. I realize we've had less than a week of it and maybe I don't know enough about it to know what I'm talking about. But as things stand – I'm almost sunk in Math. My 1st week's grade was 2.0 which is the passing mark. I devote almost all of my time to it and I still can't get it all. Assignments are long and hard – full of pure memorizing. The P's don't teach; they just grade your blackboard work and assign the next lesson. I pray that it'll become easier but the chances aren't so hot. I can't stop worrying about it.

Plebes had requirements in boxing and wrestling, half-mile timed runs, rope climbing to the gym ceiling, chin-ups, gymnastics, swimming, fencing, and even a dance class. Physical fitness was emphasized when Douglas MacArthur was superintendent of West Point from 1919 to 1922. He sought to diversify the academic curriculum with an emphasis on physical fitness and athletics. "Every cadet an athlete" was MacArthur's goal. Above the doorway to the gymnasium, MacArthur placed an inscription:

> Upon the fields of friendly strife
> Are sown the seeds
> That, upon other fields, on other days
> Will bear the fruits of victory[134]

The development of citizen-soldiers was a also goal of thirty-nine year old MacArthur, the second-youngest superintendent since Thayer, to develop an officer corps through understanding and respect rather than fear and strict regimentation. Officers were to be trained to teach, lead and inspire with a comprehensive understanding of world and national affairs.

In 1922, MacArthur was assigned to the Philippines and his successor quickly reversed many of his reforms.[135] Still, MacArthur's imprint on the academy remained, in spite of his short tenure.

If Sylvanus Thayer dominated West Point in the nineteenth century, Douglas MacArthur dominated it in the twentieth. The chief difference was that Thayer had sixteen years to impose his personality and ideas, while MacArthur had but three.[136]

"Drill & Command is over," Bud wrote to his family in March 1938, "Now we have 'Rough & tumble' or 'How to dispel mobs in three easy lessons'! We take up 'quarterstaffing' (or stick work) in the fencing room and 'how to handle man with knife' etc. in other rooms. By Act of Congress we cadets are 'gentlemen,' but they don't teach us to fight that way. The basic principle of the rough & tumble is 'give 'em the knee!' Rather crude but I suppose it's effective. However I don't intend to have 'em practice it on me."

Football and polo were the leading sports programs at West Point but the academy offered athletic programs at various levels across a smorgasbord of sports. Basketball, soccer, lacrosse, swimming, and

Love and Sacrifice

boxing pitted companies against one another, as well as inter-collegiate competitions. In his first days, Bud tried out for the Plebe polo squad, riding figure eights for the instructor, but didn't think his style of riding would suit.

"I think I did all right," he wrote, "but I fear my 'cowboyish' type of riding will get me in trouble. From all I hear they like 'form' riders."

The mid-winter horse show was held in January 1941, featuring the West Point equestrian team. Bud competed in the obstacle course race, winning the competition, but was disqualified for knocking down a stake post on one of the turns.

Bud dove headlong into intramural sports and other activities at West Point - playing polo, soccer, handball, and boxing in the 135-pound weight class, as well as joining the Dialectic Society, where he worked on lighting theatrical performances, including the One Hundredth Night Show.

During the Plebe athletic trials, Bud ran the one hundred meters in 12.2 seconds, threw a shot put twenty-eight meters, broad jumped sixteen feet, and did the highest high jump in Company G.

He wrote home, "5'3, that's almost as high as I am. I use a genuine roll which no one else does." He illustrated the letter with a drawing.

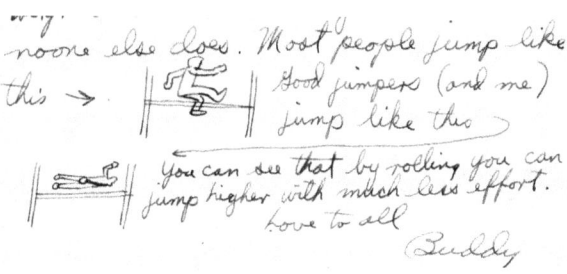

In "intramurder" basketball, as intramural sports were called at the academy, Bud's company wasn't given much of a chance, with cadet "Olly de Gruchy" singled out as "somewhat handicapped by the height factor." As an avid board game player, Bud's strongest suit at West Point was chess. In 1941, he played Isaac Kashdan, an American chess grandmaster and world champion, during a visit to the academy in which Kashdan played ten cadets and officers at once, beating them all.

Parades were a crowd favorite for visiting civilians but not necessarily as popular with the cadets. There was a full-dress parade every Sunday, starting in April, at 5:30 p.m. Daily battalion parades began in June. On Saturdays there were inspections and regimental reviews. Before football games, the cadets marched around the stadium before taking their seats, and then, from the mountains above the field, a Howitzer cannon fired the signal for kickoff. On Sundays, the Corps marched to the Plain, the wide-open area between the campus and the precipice overlooking the Hudson River, where spectators lined Visitors' Row.

A 1930 *Harvard Crimson* article, "Some 'Kaydets' Enjoy Dress Parade; Average Man Doesn't, Writes Pointer," spoofed the parade culture at the academy:

> To borrow an old definition, a parade is composed of one band, twelve hundred kaydets, and five thousand spectators. The band plays, the kaydets stand and gripe, and the spectators thrill and go home resolving to be 100 per cent Americans and vote the straight Republican tickets.[137]

Back in Missouri, Nancy Campbell was enrolled at William Jewell College in Liberty, just outside of Kansas City. From the start, she was active in the Pi Kappa Delta sorority and made the debate team in her freshman year. Her incoming mail was an item of intrigue in the campus newspaper, *The William Jewell Student*, as in this November 21, 1938 item:

> "Somebody in West Point has the mail man in a quandary when he addresses Nancy Campbell's letters to 'One on Three Melrose Hall, Liberty, Mo.' He stands and 'enie, meine, meiny, mo's' at all the little pigeon holes."

Nancy Campbell, center, with Bettie Abbie Duncan, left, and Eloise Green at William Jewell College. (Courtesy Paula Frazier)

Before Nancy left for college, Mildred threw her a going-away party. An only child with both parents deceased, no one had ever held a party in her honor.

"She says she was thrilled to pieces & had a swell time. First time anyone had a 'surprise' (she called it) party for her," Bud wrote to his mother.

With a full schedule of classes, sports, and marches around the clock, occasionally there was a chance to get away from it all:

> Ankled up to the chapel and stood on the front porch overlooking the valley for half hour. That's a rare treat I permit myself on Sunday afternoons when I'm not D. (or think I need fresh air)(or any other good excuse that salves my conscience). I like to look at the blue hills; at the little toy farms across the river; at the town north east of here; at the river; at the clouds; stars; moon. I like to think of my Mother and Father and Brother. I try to imagine what you're doing. I also wonder what that farmer over there is going out to his barn for. If the barn is warm and animal-smelly and the air is full of straw dust. I wonder what the rich fellows in the 'recuperating' establishment across the way are doing. Some ex-prize fighter has a big house where people from New York in a run-down condition (alcoholitis usually) come. He exercises and rides and hikes them all over the country until they're clear-eyed and steady-fingered again. It's a good racket 'cause next season the same people will have to come back again probably.
>
> I like to wonder who's inside that train scooting along the river bank. I think of who the most famous man on it is, and what he's doing going from Albany to New York or vice versa.
>
> It's fun to think a little – be alone a little – and pray a little.

Just like any eighteen-year-old far from home, Bud was homesick. The physical and mental stress only worsened the situation, and his faltering grades had him in near-panic of failure. In his September 1937 letter to his father about feeling academically underwater after one week's coursework, he continued:

> I try to realize that I can only do my best, and it's no use worrying if I do my best. But still the idea of failure seems to haunt me. I've never been a 'goat' before, and I don't know how to act. I hope next week that I can write you more cheerfully.

Bud did have one advantage over most cadets when it came to

homesickness and academic struggles – the support of fellow Wentworth graduates at the academy. Some knew Bud, and they all knew and respected his father. Second Classman Kenneth Griffiths was a Kansas native who graduated from Wentworth as the Honor Graduate of 1935. He kept an eye on Bud at West Point, frequently visiting him, and giving him news from home. Ted Nankivell was Bud's West Point classmate and Wentworth alumnus who had received the honor as the "most soldierly man in the organization" while serving with the Colorado National Guard before entering the academy. He tried his best to help Bud by tutoring him in math.

"Five-year man" Jay Beiser, another member of the Wentworth group, started with the class of 1939 but was "found," or failed, in French, and returned to West Point to become a beloved member of the class of 1940. As Beiser was described in a 1976 eulogy:

> As the Captain of D Company, Jack had the innate ability to deal with plebe, yearling, Second Classman, and his own classmates in such a manner that he was respected and understood.
>
> If Beiser said it, then it must be so and there was no further questioning. Really a Kansas-born oracle![138]

The Plebe year was drawing to a close, marked on March 12 by the One Hundredth Night Show, "Pass in Revue," staged by the cadets one hundred nights before graduation, celebrating the end of the "gloomy period" of winter.

"Well," Bud wrote home, "February is practically over, and Graduation is almost here. I guess I told you we celebrated 100th night the other day – about a week ago. It's surprising how accurate the sun started coming up for breakfast on that day. It hit right on the dot."

On June 13, 1938, the day before graduation, the Firsties (seniors) reviewed the remaining cadet corps in parade. Following this ceremony, the cadets retired to the Area where the Plebes were "recognized" by having their hand shaken by upperclassmen welcoming them as full cadets. Finally, the Plebes could attend a hop, the Graduation Ball.

Academically, in his first year, Bud had to tackle Tactics, Drawing, Math, French and English. He placed fifty-third in the class in Tactics, average or below in the other subjects, but was found deficient in French – meaning, he failed. The result of Bud's hearing before the Academic Board offered the chance of re-entry after a summer of remedial study followed by a test.

Bud was one of the nine cadets found deficient in French. Four were "tossed back," ejected from the academy, four (including Bud) were given the chance to pass a re-entry exam at the end of the summer, and one was reinstated after his hearing with the Academic Board.

Though bitterly disappointed, Bud resolved that he would hit the books hard and regain acceptance to the academy. If he passed, he would return to the academy in January 1939 to repeat the second half of his Plebe year as a member of the class of 1942. He wasn't the first to struggle academically at West Point. Cadet George S. Patton had failed mathematics in his Plebe year and returned to the academy the following year after a summer of study. Bud would have to follow that same course back to the academy.

★★★★

Bud spent the summer of 1938 at the Millard Preparatory School in Washington, DC, studying French and a full menu of other subjects. Millard Prep, also known as the Military Academy Prep School, was a small boarding school in the northwest section of the city run by Captain Homer B. Millard and his mother Rose. Young men attended Millard in preparation for taking the rigorous entrance exams for West Point and for remediation. Dwight Eisenhower's son, John was one of many Army officer sons who attended Millard. He wrote in 2003 that the Millard methodology was so effective that he easily won Senator Arthur Capper's appointment slot.[139]

Besides Ike, Generals George Patton, Lucius Clay, Anthony McAuliffe, Willis Crittenberger, and Henry "Hap" Arnold all enrolled their sons at Millard to prepare for West Point. Seventy-five students packed into three row houses along N Street. Classes were held in the basements six days a week and there were no electives. Those attending Millard High School Prep completed four years of education in two. Homer Millard bragged that ninety-three percent of his students passed all of the West Point entrance requirements.

Captain Millard, known by the nickname "Beanie," was something of an eccentric who made the school quite interesting and often amusing. Beanie had been accepted into West Point but was "found" and not re-admitted. Rather than being resentful toward the academy, Millard founded a school to better prepare students for the rigors of the academy's entrance exams and course work. He firmly believed that public secondary schools had become assembly lines of mediocrity, graduating students who were capable of factory jobs but not prepared for higher education.

In March 1936, Beanie and his mother, Rose, took seven students, including Bud's West Point classmate William Gordon and Yearlings Robert Strong, Page Smith, and David Crocker, and her nephew as the crew on their two-masted schooner yacht, the *Kaydet*, for a pleasure cruise to Puerto Rico, Jamaica, and Panama. They were forced to abandon the foundering yacht in storms off the North Carolina coast, resulting in their rescue by the Coast Guard. Beanie proved less than adept at sailing as he lost a second schooner, the *Cachalot*, the following year in the Chesapeake Bay.

General Crittenberger's son, Willis "Crit" Crittenberger, Jr., joined Bud in studying at Millard over the summer of 1938, along with Richard Gaspard and Karl Retzer. At the end of the course, they took their re-entry exams at Walter Reed General Hospital on August 23. Together, the four aspirants anxiously awaited the results. Three weeks later they learned that all of them had passed and would be readmitted to West Point in the class of 1942. Bud scored the highest in the group.

Their readmission would not occur until January, so three of them traveled to New York for further study with a Dr. Silverman in the Bronx, retired General Crittenberger recalled. Richard Gaspard returned to his home in Oregon where he continued his studies. Living in Dr. Silverman's home, they boned up on calculus and physics in preparation for the engineering-heavy curriculum of the second year. The group worked steadily through each weekday and spent their weekends exploring New York City.[140]

Bud and the others returned to West Point on January 2, 1939. The smoke of war was rising but still seemed very distant. In March, Germany resumed its march on Europe – first on Bohemia and

Love and Sacrifice

Moravia, within Czechoslovakia, and then Memel, Lithuania. Britain and France formed a pact with Poland to defend against aggression. Italy, under Mussolini, marched into Albania. Life at the academy was evolving with the changing world events.

Bud moved up one rung among the "runt companies," from Company G to F, and his grades markedly improved. He finished the year in the middle of the pack, at 209 in the class of 389.

President Franklin Delano Roosevelt delivered the commencement address for the class of 1939, coming to West Point from nearby Hyde Park where he had just entertained the British Royal couple, King George VI and Queen Elizabeth.

"With us the army does not stand for aggression, domination, or fear," Roosevelt said. "It has become a corps d'e'lite of highly trained men whose talent is great technical skill, whose training is highly cooperative, and whose capacity is used to defend the country with force when affairs require that force be used.

"I am sure the lessons you have learned at West Point will be of use in peace, no less than war; and that in you the Nation will take the same pride, maintain the same confidence, as, through the generations it has held for the officers of the Armies of the United States."[141]

In the heat and humidity of early August 1939, Bud's parents, along with Ted, and Nancy Campbell, visited Bud. Bud had not seen his family since leaving left Lexington in 1937. Now, they were visiting him on their way to board a ship for the Philippines. Nancy would return to William Jewel College. As a Yearling, Bud was busy with Beast Barracks and maneuvers. He took what time he was afforded to spend with his visitors. The appearance of Nancy Campbell was a surprise, indicative of Mildred's interest in keeping the flame alive.

Nancy, Ted, Bud, Mildred and Ollie
West Point, August 1939

Bud and Mildred at a West Point overlook.

Wearing his dress whites, Bud proudly showed his visitors the West Point campus – everything from the view over the Hudson River to Flirtation Walk. Whatever time he and Nancy managed together was dampened by restrictions on even holding hands on campus. A cadet found fraternizing with a femme in such a way netted punishment. Bud took everyone to Delafield Pond for a swim and game of lawn darts. The visit was brief as summer camp was coming to a close and Bud would soon be marching into the hills on maneuvers.

Signs of the military buildup were everywhere. Armored vehicles, troop transports, and jeeps clogged the narrow roads into and out of Highland Falls. Planes buzzed overhead. "Summer camp" had evolved into serious maneuvers with all of the trappings of war. Nervous excitement spread among the cadets, but the nearby civilian population was not so enamored by the mobilization. By April, West Point Commandant Brigadier General Jay Benedict had ended Sunday dress parades, which typically drew locals to the parade ground, citing "a thoughtless or malicious regard of public property." Instead, he said, cadets should observe Sunday as a day of worship and rest.[142] Bud wrote home that the neighboring towns of Newburgh and Highland Falls turned out in large numbers at football games to root against the academy team. "Familiarity breeds contempt, I guess."

3rd Cavalry Regiment (Mechanized) rumbles through Highland Falls, New York toward West Point, August 1939.

CHAPTER TWELVE

Welcome to the World of Tomorrow

Trylon and Perisphere, 1939 World's Fair. (Courtesy General Electric)

The World of Today and Yesterday will be destroyed by men who do not believe in creative humanity---by tyrants who command slaves---and no new world will be built in its place.

One side is the World of Tomorrow built by millions of free men and women, independent and interdependent...On the other side is chaos. The old ideal of men as friends and brothers becomes the Theme of today's endeavor. Working together men triumph over the forces of destruction---to build the World of Tomorrow.

- "Your World of Tomorrow"[143]

THE CAR TRIP EAST was an emotional journey for Ollie, Mildred and Ted. They would be traveling to an exotic land, for how long no one knew, as war was spreading across the globe.

Their first stop was a tour of Washington, DC and a visit with Ollie's Aunt Mabel Reed in Arlington, Virginia. Driving north toward New York, they next stopped to visit Mildred's beloved "tucco" house in Merchantville, New Jersey, and their old friends Catherine and Mac McCully. They also spent time with friends from Fort Benning, the

From left: Newton Strickland, Mildred, Reba Strickland, Henry Strickland, Newton Strickland, Jr., Ted, and Ollie Reed.

Stricklands, in New Jersey. Major Newton Harrell Strickland was stationed at the Raritan Arsenal in Edison as an ordnance officer. Their son, Newton Harrell Strickland, Jr., or Sonny, was one of Ted's best friends from Fort Benning.

Next was their visit with Bud at West Point. It was all too brief and Mildred's heart sank as they drove away. She hated leaving her Buddy, but he promised to visit them in the Philippines the next summer when he would have a lengthy summer vacation. She was used to saying goodbye to friends, but this was her Buddy.

After West Point, they visited Nellie Hollister in Connecticut. Nellie had helped bring Buddy into the world and Mildred wished so very much that Bud could have accompanied them so Nellie could see the fine young man he had become. Nellie still lived in the two-story detached home in Manchester and, at the age of seventy, she remained active in nursing.

As their departure for the Philippines was a month away, Ollie, Mildred, and Ted next went to the 1939 World's Fair in New York City. Where previous fairs had looked to the past, such as the 1893 Chicago World's Columbian Exposition that celebrated Columbus' voyage to the New World, this world's fair celebrated the future.

In what had been a marshy wasteland in eastern Queens, still known as Flushing Meadows, the 1939 World's Fair, carried the theme of "The World of Tomorrow." The fair highlighted the accomplishments of and unity among sixty countries, with the avid support of America's leading corporations. The General Motors pavilion featured Futurama where visitors sat in chairs that glided them over the United States of the future, showing wide superhighways with cars traveling at one hundred miles per hour. Futuristic houses filled planned urban complexes of modernistic designs founded in the German Bauhaus and Art Deco movements where pedestrians traversed the landscape along elevated sidewalks.[144]

Love and Sacrifice

The iconic Trylon and Perisphere rose above the fair. The two modernist structures, painted pure white, stood as the fair's "Theme Center." The 610-foot spire-shaped Trylon stood next to the Perisphere, a massive sphere measuring 180 feet in diameter. The structures were connected by a circular walkway, known as the Helicline, winding around the Perisphere to the world's longest escalator. Inside the Perisphere was a diorama called "Democracity," depicting a utopian city of the future. For a separate twenty-five-cent admission, visitors gazed upon Democracity's blending of urban, suburban, and rural areas into one unified planned community. Huge images were projected on the surface of the dome above and around. As the lights dimmed, groups of people began to march along the dome's walls and the voice of well-known radio commentator H.V. Kaltenborn boomed: "This march of men and women, singing their triumph, is the true symbol of the World of Tomorrow."

Ted Reed photo of the World's Fair.

A chorus of voices, orchestrated by André Kostelanetz, trumpeted the fair's theme song:

> We're the rising tide coming from far and wide
> Marching side by side on our way,
> For a brave new world,
> That we shall build today.

Exhausted by the years mired in the Great Depression and jittery with fears of another world war, the idea of a brighter future brought Americans hope. The consortium of industrialists behind the World's Fair saw an opportunity to present a rosy world of tomorrow driven by science and technology as the means to economic prosperity and personal freedom. New ideas and products using streamlined modern designs would ignite consumer spending, creating jobs and improving international cooperation. Where the 1893 Chicago World's Columbian Exposition celebrated the American Enlightenment by looking back 400 years to Columbus's landing on the shores of the New

World, the 1939 World's Fair displayed American leadership through industrial strength. The fair was an amalgam of ideological statement, trade show, League of Nations, amusement park, and Utopian community all rolled into one. Walt Disney recognized the potential of this themed exposition.

Global cooperation was idealized against a backdrop of international tensions. Italy and Japan, had pavilions at the fair, but Germany was conspicuously absent. Some nations put forth a strong presence at the fair in the face of tremendous distress at home. Czechoslovakia, a country no longer in existence after the 1938 Munich Agreement, wrote these prescient words on their pavilion: "After the Tempest of Wrath has Passed, the Rule of thy Country will Return to Thee, O Czech People."

Poland Participation building.

Czechoslovakia's neighbor Poland fell to Hitler's Germany in the first days of September 1939. A bronze statue of Jogaila, Grand Duke of Lithuania and later King Wladyslaw II of Poland, who defeated the Teutonic Knights in 1410, stood at the Polish pavilion as a symbol of resistance against Nazi domination.

During the time the Reeds attended the fair, the Soviet Union was quietly annexing Lithuania under the terms of the Molotov-Ribbentrop Pact signed with Germany. Russia, seeing itself as a bulwark nation, built the second-highest structure on the fairgrounds, a gigantic tower of red Karelian marble, the same used on Lenin's tomb, upon which a seventy-nine-foot statue of a glorified worker held a twelve-foot illuminated red star high above his head. To locals and fair workers, the statue was dubbed "the Bronx Express straphanger," or simply, "Joe."

Dwight Macdonald wrote about the Soviet pavilion in the July 1939 edition of *New International*: The thing that impressed me the most about the Soviet building – aside from the ugliness of its liver-red and multi-colored marble trim and the brutal heaviness of its lines – was the collection of highly dubious statements which appeared, in all the permanence of bronze and graven stone, on every wall, inside and out.

> FOR THE USSR SOCIALISM IS SOMETHING ALREADY ACHIEVED AND WON. – STALIN.
>
> THE USSR IS A SOCIALIST STATE OF WORKERS AND PEASANTS.
>
> SOCIALISM AND DEMOCRACY ARE INVINCIBLE. – STALIN.
>
> LABOR IN THE USSR IS A MATTER OF HONOR, A MATTER OF GLORY, A MATTER OF VALOR AND HEROISM.[145]

The 1939 fair would close as cold weather began to set in at the end of October and, given its popularity, plans were to reopen the next year. When it did reopen the theme no longer was "The World of Tomorrow," rather, it was changed to, "For Peace and Freedom," reflecting the spreading world war. Several nations did not revive their exhibitions, and the Soviet pavilion was demolished and replaced with an open area called the "American Common."

The seventy-five-cent price of admission to the fair is extremely modest today, but in Depression-era America, that cost was exorbitant. Fairgoers and vendors alike protested to the governing board for a reduction in the cost but without success. For the three Reeds, admission was two dollars and twenty-five cents, plus a nickel each for the twenty-minute subway ride from Manhattan. Upon entry, they paid another twenty-five cents for a guidebook. Some of the pavilions, such as Democracity, charged additional fees.

Trying to take in the sprawling geography of the fair was hard on the feet, but relief could be found on a bus for ten cents, a tractor-train for twenty-five cents, or personal guide chairs, sled-like loveseats pushed by a fair employee, costing between fifty cents and a dollar twenty-five per person. On average, lunch at the fair cost seventy-five cents and dinner ran a dollar. Ted thoroughly enjoyed the amusement section of the fairgrounds, costing an additional dollar-sixty. All together, one day at the fair set the Reeds back almost fourteen dollars. For an army major in 1939 with an annual base salary of $2,250.00, it was a great deal of money.[146]

Among the amazing pieces of technology to be seen at the 1939 fair was the television. The first television signal was broadcast to several hundred New York City homes from the fairgrounds. General Electric

 touted their developments in the new visual radio signal, as did Westinghouse, which had a studio where visitors like Ted Reed were seen on a mirrored screen in a nearby booth where his parents were watching.

Electro the Westinghouse MotoMan, a seven-foot-tall golden robot, was intriguing. A presenter slowly spoke commands into a telephone and Electro responded. "Electro…come…here…please," and suddenly Electro's motors whirred and gears spun as he stepped forward until the presenter told him to stop. Electro had twenty-six movements, a vocabulary of seven hundred words, and smoked cigarettes.

The Playground of Science displayed devices that translated human voice into wavelengths on a screen or a light responding to sound. Ascending an escalator under a twenty-five-foot swinging pendulum with an hour glass at its center, the Reeds entered a display familiar to Ollie and Mildred, "Yesterday: A World Without Electricity," from which they moved into an array of modern conveniences, such as air conditioning and automatic dishwashers. This brought home the essential theme of the 1939 World's Fair – advancements in technology brought new time-, energy-, and money-saving conveniences to the American home, increasing consumer demand, thus increasing production and creating jobs in response to the increased demand.

Lurking on the outskirts of the jubilant fair, however, was the reality of war. On September 1st, one week after signing the Molotov-Ribbentrop Pact, Hitler launched his Blitzkrieg against Poland. Three days later, France and Great Britain, allied with Poland though mutual defense treaties, declared war on Germany. As the United States stood its neutral ground, the Second World War had begun.

German Wehrmacht marching into Warsaw, Poland. (National Archives)

CHAPTER THIRTEEN

Sailing to a New Horizon

USAT US Grant

THE RECORD-BREAKING HEAT of the summer 1939 gave way to a breath of autumn in the New York air when the Reeds awoke to a cool morning on September 12. The USAT *U.S. Grant* awaited their arrival at Pier 78 in the Fifty-eighth Street Brooklyn Army Terminal complex for the six-week journey to the Philippines.[147]

As their departure approached, Mildred thought back to the time when Ollie swore that he did not want the army life for his wife but she would not have had it any other way. When someone asked her if she minded the ever-changing landscape of her life with Ollie, she replied: "I'd rather move than clean house." Yes, she missed the only home they ever owned – the "tucco" house, but the unknown of life in the Philippines was exciting. What truly did not sit easily with Mildred was being so far from Buddy. It was one thing to be across land, but soon she would be separated from her beloved first-born by the expanse of the United States and the Pacific Ocean.

An early morning cloud cover dimmed the autumnal Tuesday sunshine and a light rain began to fall as the *Grant* was tugged away from its Brooklyn mooring, the Statue of Liberty waved her torch as the three Reeds headed out to sea through the Verrazano Straits.

The USAT *U.S. Grant* was a German-built passenger steamer christened the *König Wilhelm II* in 1907, operating as part of the Hamburg-Amerika Line between Europe and South America. The ship was seized when the U.S. entered the First World War in 1917. The *König Wilhelm II* was reconditioned and converted to a troop transport renamed the USS *Madawaska*. The transport carried twelve thousand troops to Europe in ten Atlantic crossings over the course of the Great War and brought seventeen thousand soldiers home afterward. In 1922, the USS *Madawaska* was renamed the USAT *U.S. Grant*.

Their first port of call would be Cristobal, Panama, located at the Atlantic entrance to the Panama Canal.

American warships patrolled the Atlantic Ocean and P-12 biplanes flew overhead on the lookout for German U-boats. Despite U.S. neutrality in 1939, lessons were learned from the last war about the safety of passenger ships at sea. Less than two weeks earlier, on September 3rd, a German U-boat sank the SS *Athenia* in the Atlantic, killing ninety-eight passengers and nineteen crew members.

Landing safely and without incident, mules pulled the *U.S. Grant* slowly through the locks, U.S. Army, Navy, and Marines stationed at Cristobal were clearly visible along the entire length of the canal. Recent events on the world stage made Mildred nervous, but as an army wife, she knew both the excitement of new places and the threats that could lurk beneath the surface, potentially bringing harm to her loved ones. Still, war was happening across the Atlantic, not in their Pacific destination. The Chinese and Japanese were locked in battle, but the fight was largely staying inside Chinese borders.

The U.S. had maintained an ongoing presence in the Philippines since winning the island archipelago in the Spanish-American War. This was the primary foothold along the East Asian frontier, and duty rotation to the island nation was typical in one's Army career.

Passing through the Panama Canal and into Pacific Ocean, the *U.S. Grant* sailed north to San Francisco, the final stop on the mainland U.S., arriving September 29 to take on fuel and supplies.

After eleven days from San Francisco, they arrived in Honolulu, landing at Pearl Harbor. The port was crowded with warships and squadrons of P-26s flew overhead. The *U.S. Grant*'s final stop before Manila was the port village of Sumay on the island of Guam, where the American naval base is located today. The last time the *U.S Grant* visited Guam, in May 1939, it ran aground on a reef, taking two days to free the nine-ton vessel. This visit went off without a hitch.

A line of P-26 aircraft flying over Hawaii. (Reed family)

Left, Sumay, Guam. Right, Pam American Clipper in Apra Harbor, Guam. (Courtesy War in the Pacific National Historic Park, National Park Service)

During their stopover in Guam, Mildred wrote a letter to Nancy Campbell, sending it on the Pan American Airways China Clipper[148] in time for its flight back to the U.S. Nancy passed the letter along to the Lexington newspaper, which published it in full. In it, Mildred wrote glowingly of resplendent nature, "even more beautiful than the pictures one sees because you don't get all the shadows and shades of coloring in a picture," and a bit about their exotic adventures.

> The Hawaiian boys were riding surf boards, so graceful and picturesque. We decided we must try one. Can you picture us gliding blithely on the crest of the waves on a surf board? If you could have really seen us you might have been disappointed in our graceful antics – for we laid flat and paddled with our arms – but we had fun. One bucked Major Reed off and whacked him in the jaw when he laughed.
>
> Our orders have been radioed out to the ship so we know where we will be assigned, 31st Infantry, Post of Manila, Manila, P.I. There are no quarters on the post for us so we will have to rent an

apartment out in the city. It is very exciting, everyone on the boat finding out where everyone else is to go and what they know and don't know about what is in store for us. The 26th will be here before we know it. This has been such a pleasant trip we hate for it to end and yet we are eager to see what's ahead of us, too.

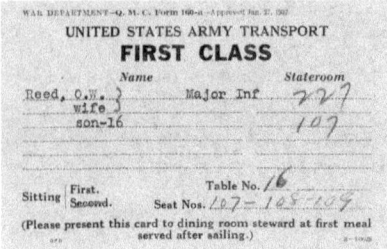

Mildred and Ollie walk to Butler's Emporium in Sumay, Guam.

Caption on back of photo Ollie sent to Bud at West Point.

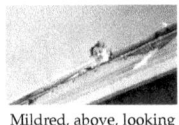

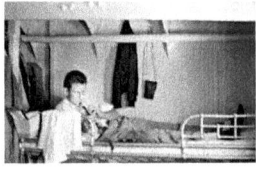

Mildred, above, looking below, while Ted plays a pennywhistle in his cabin.

Mildred and Ted wearing leis during Hawaii stop.

The *US Grant* docks at Pier 1 in Manila, Philippine Islands.

CHAPTER FOURTEEN

The Philippine Islands

THE U.S. GRANT LANDED SAFELY in Manila on October 26 to the sounds of a brass band, an event that greeted the *Grant's* passengers at every stop along the way. In Manila, the Filipino Constabulary Band, a part of their full symphonic orchestra, provided the music serenading the arrival of the army transport.

The relaxation afforded by the six-week voyage ended as soon as the Reeds arrived in Manila. Major Reed started his new job the day after landing as the S-3 Plans and Training Officer on the general staff of the 31st Infantry Regiment. The regiment was often called the "American Foreign Legion," since it had never been posted on U.S. soil. The regimental shield featured a mythical sea lion of the U.S. Army's Philippine Department, while the regimental crest was, and still is, a polar bear, recalling the regiment's role in the Siberian Expedition of 1918–20.

31st Infantry Regiment Coat of Arms Courtesy of the Army Institute of Heraldry

Major Reed's office was located in the U.S. Army headquarters at Fort Santiago, known as the Cuartel de España, housed within the southwestern corner of the old Spanish walled citadel called the Intramuros. The defensive bulwark, the walls of which were first built in the sixteenth century by the Spanish, jutted into the Pasig River at the confluence of the tidal estuary and Manila Bay.

The Reeds set up house in the Elena Apartments at 1237 M H del Pilar, apartment 4a, in the Ermita district of Manila. Ermita housed a number of government buildings, many designed by renowned architect Daniel H. Burnham, perhaps best know for his architectural splendors at the 1893 Columbian Exposition in Chicago and for his similarly designed Greek revival buildings in Washington, DC. Like Fort Benning, the City Beautiful Movement inspired the layout and design of Manila. Ermita was also home to the University of the Philippines, as well as the luxurious homes of well-heeled Spaniards and Filipinos. The Reed's fourth-floor apartment overlooked Manila Bay with an unobstructed view of verdant Military Plaza in the foreground. Cool breezes from the bay helped offset the intense Philippine heat and humidity.

A car picked up Ollie in the morning to bring him to his office and home at the end of the day, when he wasn't on maneuvers in some remote jungle. Driving along the palm-lined Dewey Boulevard, past Luneta Park and Burnham Garden, and through the walls of the Intramuros, to the far end of the compound where he was dropped off at the St. James entrance to Fort Santiago. There, he walked across a short bridge over a moat surrounding the fort and through the St. James Gate, crested with a depiction of Saint James, the patron saint of Spain, conquering the Moors in Spain. The 31st Infantry was the only all-American regiment in the

St. James Gate at Fort Santiago, Manila, 1939

Philippines, protecting U.S. interests in the islands and at the ready to respond anywhere in eastern Asia. They were at the vanguard of a general defense plan incorporating the Philippine Scouts, comprising Filipino soldiers and their American trainers, all within the U.S. Army's Philippine Department. General Douglas MacArthur had designed the defensive plan, combining U.S. and Philippine forces, augmented by a large guerrilla force in the countryside. His father, General Arthur MacArthur, led American forces in the Philippines during the Spanish-American War.

Douglas MacArthur was assigned command of the Military District of Manila in 1922 after his stint as superintendent of West Point. Eight years later, MacArthur was recalled to the United States to serve as Army Chief of Staff.[149] He returned to the Philippines in 1935 at the request of Philippine president Manuel Quezon to oversee development of a new Philippine Army. MacArthur recognized that an undersized and ill-equipped Filipino army could not adequately defend the nation against an overwhelming power such as Japan, so he devised a plan where a small corps of regular soldiers, led by officers trained in the Philippine Military Academy, would be backed by hundreds of thousands of Philippine reserve forces in the cities and countryside. In the event of an invasion, these forces would form small, mobile units operating as guerrilla forces, inflicting heavy casualties, MacArthur believed, giving any aggressor pause before striking. As an advisor to the Philippine president, MacArthur had no direct control over U.S. forces in the Philippines. Major and then Lieutenant Colonel Dwight D. Eisenhower, assistant military advisor to the Philippines and MacArthur's right-hand man, did not necessarily share in MacArthur's enthusiasm for guerrilla land defenses, convincing MacArthur to include an air component in the defense plan. MacArthur also believed that a small fleet of speedy PT boats armed with torpedoes could hold off the invading Japanese forces at the beaches. Eisenhower wondered how this could be accomplished across a Philippine archipelago comprised of more than seven thousand islands, most of them uninhabited. While Filipino guerillas operating in rugged terrain could delay an incursion, it would take months to bring in American forces from the U.S., even if sufficient troops were available. MacArthur, much to his chagrin, completely underestimated the capabilities of the

Japanese military that was already marching across China.

In 1937, President Quezon and MacArthur met with President Franklin D. Roosevelt. MacArthur requested more military aid and Quezon pressed for full Philippine independence. FDR ignored MacArthur's request and bristled at Quezon's approach.[150] Soon after, MacArthur was recalled to the U.S. by Army Chief of Staff Malin Craig, and he chose to retire from the army, but remained in Manila as a civilian advisor to Quezon until his recall into active duty in 1941.

In Manila, the Philippine legislature slashed military spending in favor of public works projects. There was a sentiment among Filipinos that the defense of the Philippine Islands was the responsibility of the United States since the U.S. regarded the Philippines as a colony, refusing to give them their full independence. With America's attention and equipment turned toward Europe, there was little of either available for the defense of the Philippines, leaving it vulnerable.

During a 1938 visit to Washington, in spite of whatever doubts he may have had, Eisenhower made the rounds asking that more military aid and attention be given to the Philippines, America's frontline against Japanese aggression. There was general agreement with Eisenhower's words, but supplies and manpower were simply not available. Eisenhower stayed in his position until December 1939 when he left the Philippines to join the 15th Infantry at Fort Ord, California.

The Philippines was regarded as a plum assignment for an army officer, but the life was anything but idle for Major Reed. As the Plans and Training Officer, Ollie was charged with implementing the defensive plans initially devised by MacArthur and adapted to the frequent revisions in the war plans, requiring regular trips to the mountainous jungles to study the terrain and preparations. These could last up to ten weeks at a time. As Mildred wrote to a friend, the pace kept Ollie's "nose to the grindstone." He also had to adapt to monkeys throwing coconuts at his tent and wild cocks crowing at dawn. One officer in his group found a python under his cot.

Initially apprehensive about the new assignment, the Reeds quickly came to love the Philippines with its varied landscapes and native cultures. Of a trip she and Ollie took around Luzon, Mildred wrote:

A trip into the provinces ("country," we'd say at home) is teeming with unusual sights. We meet carabaos ambling along, drawing two-wheeled wagons or see them in the fields pulling crude wooden plows. Toward the close of day we find them wallowing luxuriously in muddy ditches beside the road. We pass pack ponies with panniers of green bananas on each side. There is a barrio (village) every few miles.

The people love growing things and no matter how humble the hut, there are always plants in cans or cocoanut husks hanging in the window. Sway-back, big-bellied hogs wander around under the houses and in the street, and scrawny chicken little larger than bantas are everywhere.

Every stream has a dozen or so women on the banks or sitting out in the water doing their eternal washings. Their laughter and chatter sounds like a bridge party. Many of the women smoking big black cigars. We hold our breath when they come striding along the road with baskets of laundry or jugs of water on their heads, but we have yet to see one spill her load.

Ollie and Mildred cruising along a Philippine Islands river.

The sheer, full-sleeved native dress is very attractive. The skirt train is tucked into the belt at the back, disclosing a triangle of handmade lace or embroidery on the petticoat. The men usually wear white trousers and gaily-figured shirts, tails out, of course. They carry a bolo in their belts, and on Sunday, each one has his rooster under his arm; for that is the big day for cock-fights.

One rides for miles through groves of tall cocoanut trees. It is beautiful to see the sun shimmering on the palms silhouetted against the azure sky. We pass fields of sugar cane and clumps of feathery bamboo. We have seen rice in all stages. The natives stand ankle deep in muddy water and plant it to the rhythm of guitar music. They all wear something red to attract the attention of the

Goddess of Fertility, so the crop will grow well. After it is ripe, we see carabaos treading out the grain and, later, we see it drying on mats beside the road. Occasionally a woman pours it slowly from a basket to a mat to let the wind blow away the chaff.

About a hundred miles north of Manila, we left the coast and rode into the mountains where we visited the resort city of Baguio and on beyond to see the famous rice terraces that are said to have been old when Christ was on earth. At the highest point, 7,800 feet, we were not more than thirty miles as the crow flies from the ocean. The road is cutout of the mountainside. One-way traffic is controlled by gates every eight kilometers. Rock slides; cause much damage, but fortunately, we didn't come across any that we couldn't edge around. The scenery was magnificent, but Ollie made me hush oh-ing and ah-ing about it. Said he didn't enjoy scenery while driving a car where he could lean out of the window and spit seven thousand feet. We liked the little mountain people, despite the fact they are headhunters, and eat dogs.

The women wear skirts and no blouses. The men reverse things and wear shirts and no pants, only gee strings. Their lack of clothing is not offensive, however. I suppose they don't look raw like we do when we are naked because they have nice brown skin. I was highly amused at one man's apparel. He had on the usual gee string and a shirt fastened with a zipper, which struck me as a blending of the extremes in modern and primitive attire.

At West Point, when the order was shouted, "furlo class dismissed!" Bud and the other third-year "Cows" ran for the exits. This was the start of their ten-week furlo and there was no time to waste.

After a few days, particularly to see a girl he met in the spring, Bud took the train to San Francisco where he set sail on the nearly month-long Pacific crossing to visit his family in the Philippines. The train shaved two weeks from Bud's travel time in each direction. Still, he was left with only six days to spend with his family before sailing back on July 26, to return to the academy by August 28.

In November 1940, Mildred wrote to her friends about Bud's visit:

Love and Sacrifice

Bud, Mildred, Ollie and Ted in the Philippines, 1940

The highlight of the year was the six days in July that Ollie Junior spent with us. Between the second and third years at West Point students are given ten weeks vacation. We were pleased that he was willing to spend six weeks on a transport traveling to see us. Big Ollie took leave and Ted skipped school (schools begin here in June, you know), so we could all be together every minute.

Those happy days will always be a shining spot in our memories. Buddy said he was glad he came to visit, but he wouldn't want to live here. Too many queer smells!

Bud and Ted with "Pygmys" in the countryside.

They traveled by ox-drawn carts and ventured deep into the countryside. The family packed as much into their brief reunion as possible. But, in no time, Bud had to turn around and make the month-long journey back to school.

For Ted, life in the Philippines was easy going. He attended Ned Sword's Bayside Tutorial School, where the school year ran from September to March, ending early due to the tropical heat. Classes ran from 7:00 a.m. to 1:00 p.m., leaving the remainder of the afternoon for horseback riding at an airfield on the edge of the city. Ted became an accomplished rider, especially enjoying jumping with his horse, Captain Cook. He was not as interested in school as he was in horses and girls, so the schedule suited him perfectly. It was in the Philippines where Ted decided to pursue a future education in veterinary medicine.

In his 1989 Smithsonian biography interview, Ted recalled his options:

I could have joined the

Ted with his horse, Captain Cook, left, and a monitor on the right.

Medical Corps, the Dental Corps, the Chaplain Corps, or the Veterinary Corps. I had no desire to heal humans; I didn't want to stick my hands inside people's mouths; and I didn't want to preach. So, I decided I'd be a veterinarian. Of course, still at that time we had the horse cavalry, and veterinarians were an important part of that.[153]

Ted was living in a tropical wonderland where he was introduced to many new species of animals, and new and unusual habitats. In spite of the bad eyesight that kept Ted out of the regular army, he developed his photographic eye while living in the Philippines.

The man on the left is a Ilfugao tribesman near Baguio, north of Manila. The women in the middle are possibly Igorot from the mountains of Luzon. The photograph on the right was captioned: "This man has killed a man and taken his head." (Ted Reed photos)

Mildred remembered an island-hopping vacation Ted took aboard a freighter one summer with a couple of his friends:

> During vacation, he enjoyed a three weeks trip on an inter-island freighter to the southern islands, visiting Cebu, Zamboanga, Jolo, cities as fascinating as their names sound. His most thrilling experience was when he thought his soul was about to be ejected by a white-clad Moro, who advanced toward him in the gloaming, brandishing his kriss.[151] Fortunately, the man was not "juramentado" as he looked,[152] but was returning from a religious ceremony. Several times since we have been here, there have been reports of Mohammedans killing Christians in order to insure certain entrance to paradise for themselves.

During her free time, Mildred busied herself sewing and knitting for the Red Cross, working at the Y.W.C.A., teaching Sunday school, and attending bible classes. Ollie picked up golf, while Mildred played volleyball and badminton, "for my 'figger's' sake." Attending and hosting various events kept Mildred and Ollie busy as a couple.

Young Moro man on the Zamboanga Peninsula, Mindanao (Ted Reed photo)

Open-air store in Manila.

Mildred immersed herself in Philippine culture and the surrounding world, more than happy to vividly describe the adventures of shopping in Manila to folks back home:

> Picture the stores on Main Street at home all open doorway instead of show windows. You see all their merchandise without getting out of your car. The ambitious clerk here invites you in as you pass, but the majority of them are quite languid about waiting on you even when you have found what you want.
>
> It is hard for me to become accustomed to the fluidity of prices. They ask several times what they expect to get, and in order to save face, you must haggle half an hour until the price becomes reasonable.

Philippine market.

Market stand with an armed man.

The Chinese stores with their ivories, linen, silks and laces are the most tempting places. The shops selling native carvings and baskets awaken all my acquisitive instincts. We like to spend an afternoon poking around in a street called "Thieves Row." The shops are tiny rooms, mere holes in the wall, piled high with all kinds of junk that is supposedly purchased from robbers.

Everything is covered with dirt. You have to scrape it off to see if a goblet is tin or silver. There is no semblance of order. You may have to untangle a horse bridle to get to a hurricane lamp. It is very exciting because you never know when you may unearth a treasure. I haven't yet, but I hear it has been done! But the food market is the place to go for real atmosphere! Here you can buy fish, dead or alive; eggs, fresh or rotten; vegetables, green or decayed. Here's freshly butchered meat hanging in open stalls, a Roman holiday for flies.

Screens, you ask? Don't be a sissy! Watch the butcher hack off a hunk for that customer and wrap it in an old newspaper. Such a hub-bub! Women arguing over prices. Gutter snipes quarrelling over the contents of the garbage cans. Squealing urchins and mangey dogs darting about underfoot. Ducks, turkeys, chickens add to the commotion. And oh, the smells arising on the warm tropic air! Whew! You swear you will never again be able to countenance the sight of food, and then as you go on and things get thicker, you wonder if the food you ate for breakfast is going to behave properly. At last, you have found your way to the street and can breathe once more. In a week or two you are amazed to hear yourself saying to a newcomer, "My dear! Have you been to the Market? You must not miss it. Let me call for you in the morning."

Are you worn out with all this description? But we won't let you go until we have taken you for a stroll along Dewey Boulevard. This pleasant tree-lined street follows the shore line of the Bay for four miles. On one side, rambling old Spanish houses are half hidden by high stone walls over which purple bougainvillea climbs. Through the iron gateways we have a glimpse of the gardens, cannas, hibiscus, cup-of-gold

Notice all the strange trees (What wouldn't I give to see a cottonwood! Prettier than any of these tropical trees, if you ask my candid opinion). There are palms of all kinds but my favorite is the flat one that looks just like a giant palm leaf fan. That immodest tree with its roots exposed is the banyan. The flame tree is the shape of an umbrella, and when it is covered with red flowers, it is gorgeous. The queer dead-looking tree with bare limbs sticking out at right angles from the trunk is, guess what? A kapok tree. Has pods like a milk weed full of fuzz.

Postcard showing a carratela beneath a flame tree.

The people one meets along the boulevard are as unusual as the trees. Here comes a Hindu with his head in a white turban. His silky black beard intrigues me. I think I will try to induce one of my men-folks to grow one like it. Many firms and apartments hire these awe inspiring Indians as night watchmen. Occasionally, we see a Hindu woman in her picturesque sarong. There is one who has a diamond on the side of her nostril. Has a hole pierced like we have for earrings. We always meet a group of brown-robed priests from a monastery inside the old walled city, taking their constitutional. They wear barefoot sandals and have a round shaved spot on top of their heads.

We always see, too, jaunty American sailors on shore leave. Fresh, clean-looking lads. We are proud of them. Here comes a Moro offering to sell us pearls. The Moros wear the red or purple fez of the Mohammedan. They come from the southern islands in their gay-sailed vintas. In the early days they came up here to kidnap and plunder, and the natives of this island still look upon them with wary eyes.

We pass many attractive Filipino young people looking neat and trim in American style clothing. There are many colleges here in Manila where they congregate, eager for an education. The boulevard is a favorite place for nurses to bring their perambulators and toddlers for an airing. We especially like to watch the Chinese amahs, so attentive to their small blond charges, many of them Navy youngsters. The amahs wear high collared long middies, shiny black sateen britches and flat satin slippers. Here's a man who insists on us buying a sweepstakes ticket. "Segee. Segee." That means "Go away" in Tagalog.

Mildred's friend speaking with a street vendor.

The vendor is an ever-present and interesting figure with his big flat hat, flapping shirt tail, short trousers, jogging along with a bamboo pole over his shoulders. From each end of the pole hangs a basket, carrying his wares - little cakes, peanuts, dress goods, flowers, puppies, ice cream, whatever he has for sale today.

Our attention is attracted by the clitter-clatter of pony hoofs coming down the street and the jingle of the brass ornaments on his harness. We see the calesa, the most universally-used means of native conveyance in the city. It is a two-wheeled buggy, painted green, black or red, trimmed in brightly painted flowers resembling examples of Swedish decoration I have seen. On each side are brass carriage lamps. The driver sits on a

A calesa outside Manila Cathedral.

stool almost against the dash board, leaving room for two or three in the seat. The seats are often covered with matting and we have been warned that certain unmentionable insects find the calesa a convenient place to change customers. We do not speak from experience. Yes, I'm knocking wood.

Akin to the calesa is the carretela, a country cousin, so to speak. It is a two wheeled pony cart also, but not so ornate. Two boards are used for seats with space beneath them and on top of the vehicle for what have you? Your basket of garden produce, your bundle of firewood, your bag of laundry, your pig going to market, a stray piece of furniture and always, a few surprised chicken heads poked through holes in a basket, probably gazing their last on this world of weal and woe. And while the calesa carries two passengers, the whole family with dogs and grandchildren pile into the carretela until your heart aches for the poor pony, who must pull the load. Have I neglected to say that one drives on the left of the street over here in the manner of the British? Why we should do as Britains when we are not in Britannia, I cannot say."

Mildred, Bud, and Ted in a carabao-drawn kariton.

The Reeds shrugged off the Manila heat as nothing compared to Kansas summers, especially from their balcony overlooking the bay - but the rain was another matter. "It pours for weeks at a stretch and is so muggy and dank," Mildred wrote, "we would shout for joy to see a good old Kansas dust storm blow up. One day's rain would insure a wheat crop if it fell in Kansas."

Part of the prestigious lifestyle of an Army officer in Manila was having servants. The Reed's primary houseboy, as he was called, was Francisco, a husband and father of five. Mildred wrote to her friends of the family's home life:

We have a Filipino who cooks and cleans for us, and a lavandera who washes every day. That sounds like we must be very dirty

Francisco and his family. Last name, and fate, unknown.

people, but we can only wear a piece of clothing once, on account of perspiration. Ollie alone has three changes of clothes every day - his uniform, his golf togs and his dinner suit. Ted has his school clothes, his riding outfit and his home clothes. All the lavanderas for this apartment house sit on the floor of a shed back of the apartment. Each has a flat tin tub and she beats the clothes with a stick. Soap is some kind of a root. They iron sitting on the floor too, with clumsy charcoal burning affairs. Sometimes, sparks fly out and tiny holes decorate our clothing. We don't let it upset us. It's part of the game of living over here. The girls sound like they are having a good time, chatting and giggling.

The natives eat rice every meal and quantities of fish, raw and cooked. We buy ten pounds of rice a week for our two servants. Garlic is the favorite Filipino condiment, but those who work for Americans have to choose between their love of garlic and their jobs. The majority of cooks are men. They delight in fancy surprises. One evening when we had guests for dinner, our Francisco brought in mashed potatoes molded in the shape of a duck. He always makes artistic cake frostings and designs on the pie meringue.

The family was amused at an interview I had with him. Some ladies were coming for lunch, and I went into minute detail about the menu, recipes to be used, etc. He listened without a glimmer of intelligence until I had finished, then he said, "But, Mum! I am planning to make Country Captain. I made it for a General once and he took two helpings. It is very good, Mum." I inquired as to ingredients. Chicken, raisins, peppers, almonds. It sounded

Mildred preparing to entertain at home.

surprising, but what could a poor woman do in the face of such male determination? [Country Captain was Patton's favorite dish.]

A popular native delicacy is the "balut." It is a duck egg that has been partially incubated. You can buy a nine, sixteen or twenty day eggs as your taste dictates. They peddle them at ball-games and races like we do popcorn at home. The idea of balut horrified me. I've been real vocal about the repugnant idea. One day, coming home from a cocktail party, I said, "Did you taste those delicious little hors'derves Edna had? I must have eaten a dozen." Ollie doubled up laughing. "Those, my darling," he said, "were balut."

Ollie drew a crowd when he had to change a flat tire in the countryside.

We have found the people very courteous and friendly, especially the ones we meet out of the city. They are as interested in us as we are in them. Wherever we stop the car to consult a roadmap or eat a sandwich (we carry food and water when we are away from home), a crowd gathers as if by magic and solemnly watches us. Not long ago, Ollie left the car to inquire directions. A group of children gathered around me. They were very shy at first, but after I had grinned at them several times, they went into gales of mirth. I asked the oldest girl, "What are you laughing at? My funny hat?" She said, "Oh no, Mum. Your funny face." So, you see, they have a genuine sense of humor.

Mildred and her car were a curiosity in their travels.

The mortality rate among them is very high. They estimate 93 die every day from T.B. alone. They seem to accept illness as God's will and don't try to do much about it. Sunday seems to be the favored day for funerals. Everybody walks behind the hearse, or sometimes, the coffin is on a two-wheeled cart, decorated with artificial flowers, pulled by hand. The procession is headed by a band. The pieces they play would not be our choice for funeral music. We have heard "Yes Sir, That's My Baby," "Should Old Acquaintance Be Forgot," "I'll Be Glad When You Are Dead, You Rascal, You."

Most of all, Mildred enjoyed sitting on their fourth-floor porch overlooking Manila Bay and "the changing lights and shadow, the boats that come and go."

Mildred concluded her letter to friends:

In the shallows beyond the club, the rusting hulks of the Spanish fleet are marooned, an effective monument to the adequacy of the American Navy. As children we recited:

"Oh, dewey was the morning upon the first of May and Dewey was the Admiral down in Manila Bay" little dreaming that we would someday see the results of that encounter. Sunset hour has arrived and the sun is sinking behind mauve Mount Mariveles across the Bay, painting the sky with vivid shades of orange, then pink, then purple. It is said that in no place in the world do the beauteous sunsets excel Manila and to see one is to believe it. The old Spaniards aptly called the Philippines "las islas del Poniente," the sunset isles.

Nevertheless, the finest sunset of all will be the one we see over the taff rail of a departing Army transport about a year from now, we hope, we hope, we hope!

Sunset over Manila Bay from the Reed's fourth floor apartment.

CHAPTER FIFTEEN

War Closes In

Japanese forces in China, 1937. (National Archives)

MILDRED RECALLED DRIVING WITH OLLIE through the Manila slums, thinking of their future: "We have had such a full happy life. When we retire, let's go someplace and try to bring happiness into lives of people who haven't been as fortunate as we have been."

In her November 1940 letter to friends in the U.S., Mildred conveyed the anxieties of the times, especially being so far from home:

Dear Friends,

Christmas time approaches and thoughts of other days and far away friends are uppermost in our minds. We won't say "Merry Christmas" because who could be merry with all the tragedies and confusion of a war torn world bewildering one; but we are hoping there is peace in your hearts and faith in happier times ahead.

By the middle of 1940, military planners expected Japan to send one-hundred thousand troops to capture Manila and eradicate the U.S. naval base at Subic Bay. The Japanese were making no secret of their imperialist ambitions throughout eastern Asia, as well as their intense anti-Western sentiments. U.S. forces foresaw that such as attack would come without warning, probably as soon as war was declared, or during the dry season in December or January. These assumptions set the stage for the Army and Navy's planning, known as War Plan *ORANGE*. Initially devised in 1919 and 1924, and revised several more times over the years, *ORANGE* was renamed War Plan *RAINBOW* in 1938, setting the foundation of 1940–41 defensive planning. *RAINBOW* foresaw an alliance with England and France in a war against Germany, Japan and Italy. In the Philippines, the focus centered on the defense of Manila and the naval station at Subic Bay, but MacArthur disagreed with the assumption that the remainder of the Philippines, Wake Island, and Guam would all be lost in the initial fighting. MacArthur bristled at what he thought was defensive and defeatist thinking, but he had retired from active duty in 1937. In spite of his position as advisor to Philippine president Quezon, MacArthur had no say in American military matters.

In a September 16, 1940 response to a question posed by the Committee to Defend America by Aiding the Allies, MacArthur issued his "too late" statement. "The history of failure in war can almost be summed up in two words: too late...The greatest strategic mistake in all history will be made if America fails to recognize the vital moment, if she permits again the writing of that fatal epitaph: too late.."

He returned to active service only months before the Japanese invasion, holding onto the belief that the best strategy relied upon an aggressive defense of the Philippine Islands, combining U.S. and Philippine forces with the support of an armed citizenry to fend off any offensive actions by the Japanese. While MacArthur's enthusiasm may have been infectious, the reality was that there was little, militarily, that could stave off a major attack.[154]

The 31st Regiment had witnessed the brutality of the Japanese in 1932 during their assault on Shanghai. The Japanese had been making raids into China since 1931 seeking dominion over Chinese raw materials and labor resources. The 31st was dispatched to Shanghai in a neutral role strictly to guard Americans living in the International Settlement.

In response to the killing of Japanese citizens by Chinese Nationalists in Shanghai, Japan sent warships up the Huangpu River through the city's center shelling fortifications and sending marines ashore to seize the central business district, effectively bringing China's largest city and the Yangtze River delta under Japanese control. This marked the first battle of what would become the Second Sino-Japanese War, lasting the duration of World War II.

Internally within China, Nationalists and Communists were at war with one another but they temporarily set aside their differences to fight off the Japanese aggression. As the fighting intensified it drew closer to Shanghai's International Settlement. On January 31, 1932, then-Army Chief of Staff MacArthur issued the order for the deployment of the 31st Regiment to Shanghai.

As neutral guards, the American soldiers witnessed the brutality wrought upon all sides in the conflict but could not act. A fragile alliance between the Nationalists and Communists fractured as each side executed members of the other under suspicion of collaboration with the Japanese. Within their own ranks, members were executed as spies for the other side, as well as for the Japanese. All the while, Japanese troops were publicly beheading Chinese soldiers and citizens.

By the summer of 1932, the situation in Shanghai had calmed enough for the 31st to return to Manila. In the Japanese-controlled city, the International Settlement was an island of relative serenity amid the chaos. Before leaving Shanghai, officers of the 31st purchased a silver punch bowl and twenty-nine ornamental cups to commemorate their mission. This became known as the Shanghai Bowl and today is a centerpiece in the regimental headquarters at Fort Drum, New York.

The Filipinos had a rather relaxed attitude about the prospect of war, refusing "to succumb to prevailing war hysteria," as Philippine writer Vicente Albano Pacis wrote in the *Washington Post* on March 24, 1941. They believed the presence of American forces would dissuade the Japanese from attacking and that the Japanese would not want to get mired in a second front as they already were having trouble in China. Plus, Soviet Russia could easily attack them from the rear. The Japanese trumped that line of thought by advancing their fight with China into Indo-China and signing a neutrality pact with Stalin. Little stood in their way.

In spite of looming war, recently-promoted Lieutenant Colonel

Reed enjoyed golf at the Wack-Wack Golf and Country Club with Lieutenant Colonel Henry G. Sebastian and Lieutenant Colonel Malcolm Fortier. Another day would be spent deep-sea fishing or mountain climbing with Colonel Albert M. Jones, commanding officer of the 31st Regiment, and S-2 officer, Lieutenant Colonel Irvin E. Doane. In the days leading to Mildred's departure, Ollie and Mildred attended a polo match in the afternoon with Lieutenant Colonel Leonard R. Crews and his wife Carmen, followed by dinner at the jai-alai games where they were joined by Major Glen R. Townsend, wife Edith, and daughter Virginia, who had crossed the Pacific with the Reeds and Crews aboard the USAT *Grant*.[155]

"Cost us a pretty penny," Mildred wrote to Ted, "but I told dad we would count it as our silver wedding celebration, because we will probably not be together in August."

Americans in the Philippines clearly knew that war was coming, and in an interview with Pamela Henson of the Smithsonian Institution, Ted Reed recalled his father's concerns:

> [He said,] Ted, we're going to be at war. I want you to go home and you can join if you want to, do anything you want, but I want you to have your training. I don't want to send you to the front line. That's what's liable to happen. We're liable to gather all American citizens, give them a rifle and say, GO.[156]

Army officials prepared to evacuate all U.S. civilians by July 15, then moved the date up to May 15, 1941. The Navy had already evacuated its civilians. Japan was fortifying their League of Nations-mandated "protectorates" of the Marianas and Caroline Islands, harassing sea traffic between Hawaii and the Philippines, as well as holding the island of Formosa (now Taiwan), just sixty-five miles from the northernmost island in the Philippine chain. In July, they established bases in French Indochina (now Vietnam) to the west of the Philippines.[157]

In March, Ted left Manila's Pier 3 on the USAT *U.S. Grant*, the same ship that brought the Reeds to Manila, bound for San Francisco. As he stood along the rail of the departing ship, Ted felt silly holding the balloon his mother had given him so she could spot him among the other passengers. He managed one photograph of his own as the ship pulled away. Among the throng crowded along the pier, Mildred and Ollie stood at the very end waving goodbye.

Ted, left, holding the balloon his mother had given him.

The USAT *Republic* was scheduled to sail next. It had arrived in Manila on April 22 with two thousand Army reinforcements, equipment, and supplies, all of which was immediately ferried across Manila Bay to Corregidor, an army stronghold known as the "Philippine Gibraltar." The *Republic* would be used to evacuate civilians back to the U.S., but Mildred did not want to leave Ollie until the last possible second. She wrote to Ted soon after his departure, "since it [the Republic] is a nice boat [we hope] that enough people will volunteer to go on it that we will not be ordered." In spite of her desire to stay, Mildred was sent home aboard the *Republic*, along with eight hundred women and children, leaving her husband behind.

Mildred Reed and Mae Murphey in Honolulu.

Mildred joined Ted in San Francisco, where he was completing his senior year at a prep school. From there, they moved to Manhattan, Kansas, to begin studies at Kansas State. A number of women with husbands in the Philippines were living in Manhattan. They regularly exchanged their news from across the Pacific. Mildred recalled one:

> I was with Edith Townsend, one day, when the mail came. She started reading a letter from her husband, aloud. He was telling how desperately ill Ollie was. She tried to stop reading but I said, "Go on. That was written three weeks ago, I'd have been notified if he were dead, so he must be better."

Undated scene from Mildred Reed's Philippines album, presumably shot by Ted Reed.

CHAPTER SIXTEEN

Laura

Laura Sloman and Cadet Bud Reed, West Point, February 1941.

WITH BUD LIVING over twelve-hundred miles away at West Point, Nancy Campbell and her West Point beau remained the subject of gossip at William Jewell College. She and Bud continued to be very low-key about their relationship, so rumors continued to circulate in the campus newspaper.

The whispers about Nancy and her West Point flame started in 1938 and continued into the next year when the gossip column in *The Student* newspaper at William Jewell noted, among a list of engagements taking place over the Christmas holidays, "Nancy Campbell started wearing that Army pin!" And, in another column, "About Most Anyone," noted: "Nancy Campbell is wearing the most interesting West Point 'Miniature Ring.'" A West Point cadet giving his girl a miniature ring was generally taken as a gesture of engagement.

As the star of the college debate team and a fixture on the honor roll, Nancy stayed busy at college, but by the spring of 1940, she was feeling a distinct chill from Bud. As noted in the May 13, 1940, *The Student*, Nancy began finding her own way: "The tall, handsome Harold Poynter seems to be getting along with Nancy Campbell right well."

One week later, there appeared to have been closure: "Nancy Campbell rates the No. 1 story of the week. She got a long distance call from her West Point swain and guess where they found her – under the water-tower and not alone."

With their relationship icing over, Bud fell head over heels for sixteen-year-old Laura Sloman who had accompanied her older sister Margaret on a trip to the academy from their home in East Orange, New Jersey. Soon after meeting Laura in the spring of 1940, Bud sent her a copy of the *Week End Pointer*, a guide for young ladies, or "femmes," visiting West Point cadets.

Bud wrote on the opening page: "To Laura with anticipation of many happy weekends together from Bud." Below this was the printed dedication: "Dedicated to THE FEMMES. For what would a weekend be without them?"

Week End Pointer Bud sent to Laura, Spring 1940

The *Pointer* covered everything Laura needed to know about visiting her cadet - from train schedules along the West Shore Line between Weehawken, New Jersey and West Point, as well as accommodations, recommending the Thayer-West Point Inn, a hotel still in operation. Since the Thayer is located on the West Point campus, cadets were allowed to visit their femmes at the hotel without leaving the campus. The Thayer even set aside a wing for the special purpose of housing young women in a dormitory-like setting. The booklet also introduced visitors to Flirtation Walk, where cadets and their female visitors could "walk and talk intimately."

"Many, many people," the guide noted, "have found consolation and relaxation there, and in the present day, it is perhaps the only place where cadets may truly 'fall out.'"

Even today, cadets are not permitted to hold hands with their visitors while on the campus, and Flirtation Walk remains a hideaway, closed to outsiders.

Laura and Bud at West Point, February 22, 1941

Bud fell instantly in love with the shy teenager, five years his junior. They were kindred spirits who shared more than a physical attraction to one another.

In the first sixteen years of her life, Laura moved multiple times, with her mother and father, Mike and Catherine Sloman, along with her three older siblings, Christopher, Margaret and James. They alternated between luxury and poverty, owing to the Great Depression and Mike's business fortunes. One relative looked back speculating that Mike Sloman always stayed one step ahead of the law.

Morris and Richard Sloman, 1906

Born in 1894, Morris Sloman, later known as Michael, grew up in an advertising family in Dayton, Ohio. He briefly worked in the family business assisting with major corporate accounts but moved to Columbus to strike out on his own. There, in 1912, he married Blanche Griffith, and in 1919 Blanche gave birth to a son, James. They had lost one son in childbirth two years earlier. Young Morris quickly became an esteemed ad man who was recruited to join George Creel's U.S. Committee on Public Information, promoting the causes of World War I to the American people through marketing and advertising, what would become known as a propaganda campaign. He built upon his accomplishments in communications and became a principal in James M. Cox's 1916 campaign for governor of Ohio. In 1920, Morris's name came up in U.S. Senate hearings into campaign financing from his activities with the Forward Looking Association and their "Let's Redeem Ohio" campaign on behalf of Cox. A U.S. Senate investigation looked into charges that $37,000 raised in 1916 presumably for a Dayton, Ohio flood control fund had been laundered into the Cox campaign. Dayton businessman Adam Schantz contributed $7,000 to he fund and raised an additional $32,000 from five other wealthy Dayton businessmen. Schantz was reported to have deposited the money into his personal account and then dispersed the funds to Morris Sloman and two others.

"I would say I made, if not all, about all of the payments to Mr. Hays and Mr. Sloman and Mr. Burba," Schantz told the subcommittee.

"Who told you to give such large sums as these to an association you knew absolutely nothing about? Senator Walter Edge, a Republican from New Jersey, asked Schantz.

"Oh, I had confidence in Mr. Hays, and had more confidence in Mr. Sloman, and I had confidence in Mr. Burba."

Morris did not appear before the committee and nothing became of their investigation, in spite of the subcommittee chairman, Iowa Republican Senator William S. Kenyon, barnstorming the hearings through several Midwestern states to discredit the Democratic candidate in the 1920 Presidential election - Governor Cox and his vice-presidential partner, Franklin Delano Roosevelt. Republican Warren G. Harding prevailed.[158]

By the time of the hearings, Morris, Blanche, and infant James had left Ohio, and were living in a crowded boarding house outside of Pittsburgh. Morris was working in advertising sales for local newspapers, but this failed to satisfy Blanche. She left him in 1923 and moved to Cleveland, where she remarried British-born ad man, William Young. James was placed in the care of Morris's sister, Helen Sloman Pryor.

Laura Gillmore

Morris married Laura Gillmore, a teacher and education pioneer, on September 20, 1923, in Erie, Pennsylvania. Laura was born in Moravia, New York, educated at Columbia University Teachers College, and taught at the innovative Moraine Park School in Dayton before becoming supervisor of Cleveland elementary schools in

1921. Morris was listed on the marriage license as a "publisher" from Pittsburgh.

After the Erie wedding in a Unitarian church, the couple moved into Morris's Pittsburgh home at 4181 Centre Avenue. Laura stopped full-time work in schools devoting her time to writing a book about education principles. On December 30, 1924, Laura gave birth to a daughter, Laura Gillmore Sloman.

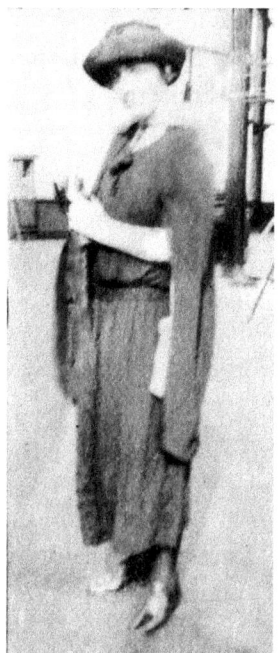

Laura Gillmore

Infant Laura Sloman with Aunt Helen Pryor.

Four days later, Morris's wife died of complications from a pre-existing kidney condition known as Bright's Disease. She was buried in an unmarked pauper's grave in Pittsburgh's Allegheny Cemetery. Infant Laura joined her older step-brother James in the care of their aunt, Helen, for the next eighteen months.

Posthumously published in 1925, the first edition of Laura Gillmore Sloman's *Some Primary Methods,* was regarded for many years as an important guide for young women approaching a career in elementary education.

Photoplay, June 1928
(Media History Digital Library)

With Laura's death, Morris went into a tailspin while continuing to live at the address he shared with her. He moved on from newspaper advertising to president of Fayro Laboratories in Pittsburgh. Fayro offered the "Fayro Hot Springs Home Bath Treatment," a seemingly amazing weight loss method achieved through soaking in natural mineral bath salts drawn from "the waters of 22 hot springs of America, England and Continental Europe." Fayro targeted celebrity magazine readers. A 1928 ad in *Photoplay* magazine read: "Fayro, by opening your pores and stimulating perspiration, forces lazy body cells to sweat out surplus fat and bodily poisons. Add Fayro to your bath at night and immediately you will lose from 2 to 4 pounds in an easy, refreshing and absolutely harmless manner."

In 1926, Morris moved to Chicago for a job in the advertising department of the *Chicago Herald & Examiner* newspaper. By 1931, Fayro was being pursued by multiple federal and state authorities. In September 1932, the advertising firm Hampton, Weeks & Marston filed suit against Fayro for $16,577, and, in February 1933, the company finally cleared bankruptcy in New York, with offices located at 354 Broadway, as Fayro Sales Co. with Maurice H. Sloman and William M. Marner at the helm. Perhaps the most damning charge against Fayro and their amazing weight-loss bath salts came from the Federal Trade Commission. Citing Fayro's claim that, "'when used in a hot bath, without any ill effects, being beneficial in the earliest stages of Bright's disease, and recommended by doctors', when in truth Fayro is not such a concentrate as is described, and will not effect reduction as advertised and may be highly injurious when used as advised in the early stages of Bright's disease." This was the disease that killed Laura.

Morris and Katherine Sloman, undated.

On June 11, 1926, Morris returned to Pittsburgh from Chicago to marry Jane Katherine Scaife, a divorcée with two children, Christopher and Margaret. Katherine, who later changed her name to Catherine, had gained custody of the children after filing for divorce on the grounds that her husband, Christopher T. Scaife, had abandoned his family for over two years. There was talk that Christopher was homosexual.

The new Sloman family settled in the Chicago suburb of Oak Park, where Ernest Hemingway spent the first twenty years of his life, and a community where there are more Frank Lloyd Wright-designed houses and buildings than anywhere else in the world. The Sloman family lived in Oak Park for the next four years while Morris held a post as advertising director at the *Herald & Examiner*, a Hearst newspaper. The family summered on Pistakee Bay, a little over fifty miles northwest of Chicago. Five-year-old James and eighteen-month-old Laura joined their father and his new wife after living with their aunt. But when the Great Depression struck, Morris was one of millions put out of work, and the family was again on the move.

Christopher, Morris, Katherine, Laura, Margaret, and James Sloman, undated.

In 1930, the Slomans arrived in Cleveland. Morris assumed the job of vice president and treasurer with Sloman-MacKinnon, Inc., a company selling refrigerators and electrical supplies that incorporated and folded in the same year.

From Cleveland, the Slomans moved to Westchester County, New York, in the suburbs north of New York City, living in a nine-room, three-bath Tudor-styled home at the corner of Glenwood Road and Edgemont Place in Edgemont, often regarded as a neighborhood of Scarsdale. There, Christopher and Margaret acted in seventh and eighth grade plays at the Edgemont School, and Laura won a prize for her costume in the six-and-under group at the 1931 Ice Carnival.

After the 1932 school year, it appears Morris and Katherine separated. She moved to the nearby town of New Rochelle, while Morris moved to the Bronx. It was likely during this time when Laura and her half-brother Jim recalled seeing their father selling apples on the streets of New York.

Margaret lived with her uncle, Richard Sloman, and his wife, Mabel, while attending New Rochelle High School. During the summer of 1935, she moved in with her mother when Mabel Sloman left to take her daughter, Mabel Jane Sloman, to Hollywood. Mabel Sloman convinced Rudy Vallee's agent, Morton Millman, to let her daughter perform in January 1935, and the child was an instant hit when

Ann Gillis as Becky Thatcher

she stole the show from Vallee at Manhattan's Hollywood Restaurant. The red-headed eight-year-old brought the house down when she stopped to pat the head of "Red" Stanley, leader of Vallee's orchestra, the Connecticut Yankees, asking, "Where did you get my hair?" From there, young Mabel's mother took her West in hopes of becoming the next Shirley Temple. In Hollywood, she became Ann Gillis, perhaps best known as Becky Thatcher in the 1938 film, *The Adventures of Tom Sawyer*.

Living with her mother, Margaret graduated from Yonkers High School in 1936. Morris's stepson, Christopher, graduated from Dewitt Clinton High School in the Bronx where he ran track and his stepbrother, James, graduated from Theodore Roosevelt High School, also in the Bronx. James attended the City College of New York and joined the Citizens Military Training Corps on the campus in 1936. He enlisted in the Army Air Corps the day after Pearl Harbor, rising to the rank of captain in command of a squadron in Palm Springs, California during the war, and at the Abadan Air Base in Iran post-war.

In New York, Morris worked as an assistant to famed columnist Arthur Brisbane. Morris met Brisbane while working at the Chicago *Herald & Examiner*, where Brisbane was the editor, columnist, and close confidant of William Randolph Hearst. Brisbane had become entwined with the Hearst newspaper empire, entering into joint business ventures with the tycoon, such as Hearst-Brisbane Properties, investing in New York real estate and buying newspapers on behalf of the Hearst Corporation. Following Brisbane's 1936 death, Morris, now known as Mike, found employment selling furniture advertising at the *New York Post*. The gregarious Mike Sloman was often referenced in *Post* society items preparing cocktails, fishing, or playing golf.

Mike Sloman's fortunes appeared to be on the rise when he signed the lease on an apartment at 180 Riverside Drive, a building overlooking the Hudson River, in 1937. The family moved to East Orange, New Jersey in 1938, while spending time at the Riverside Drive apartment, as Ann Barbour, Catherine's niece, recalled visiting them at the apartment during a trip to the 1939 World's Fair. By this time, Katherine had changed the spelling of her name to Catherine.

The constant moves led to an unsettled life and the one who

suffered most was Laura. Catherine, the woman she was raised to believe was her mother, was cruel to her, often calling her "stupid" and "retarded." Laura stopped attending school after seventh grade, in 1936. While the rest of her family had at least completed high school or more, Laura was something of Cinderella in the Sloman household, suffering constant indignities, leaving her extremely shy and reticent. It would not be until her adult years when Laura would learn the truth about her birth mother and namesake, Laura Gillmore Sloman.

Laura's older step-sister, Margaret, spent time with her, taking swimming lessons under the instruction of Gertrude Ederle, the first woman to swim across the English Channel, at the Playland Park in Rye, New York. As an adult, Margaret was quite attractive, and stories from the family say she appeared in the *Saturday Evening Post* and was a stand-in for the actress, Hedy Lamarr[159]. In the spring of 1940, Margaret brought her little sister Laura along on a trip to West Point where she met handsome Cadet Bud Reed.

Margaret, left, and Laura Sloman with Gertrude Ederle, the first woman to swim the English Channel, 1932.

Laura and Bud were both very shy, a result of growing up without regular friends, as their families were constantly on the move. Although they were nearly five years apart in age, the pair of shy kids from tumultuous lives fell into one another's arms.

Nancy faded from Bud's mind as he lovingly gazed upon this thin, shy, and quiet big-city girl. The constant moves throughout childhood they shared developed similarities in character, but there were significant differences in their upbringings. In spite of all the places Bud lived, his roots were firmly planted in small-town America. As uprooted as his life had been, made worse by his father's long absences, he always felt the deep love of his two parents, particularly his mother. Laura, on the other hand, lived a stressful life; her father, whom she dearly loved, struggled against the economic tides to provide his

Laura Sloman in Beverly, NJ, undated.

family a decent life, while her stepmother treated her cruelly. Laura latched onto Bud as quickly as his arms reached out to her.

"Your very sweet letter came tonight," Bud wrote to Laura in 1940, "and makes me feel happy – although not so happy as I was two nights ago. You are right – we do have a lot of fun together. I hope we always will."

Laura had fallen in love with her handsome West Point cadet, just like in the movies. Soon, she was regularly taking the train to West Point for football games and dances. They attended the Notre Dame game at Yankee Stadium and the Army-Navy game in Philadelphia. As she remembered in her March 22, 1942, diary entry, "By the way a year ago today I got my 'A' pin," her army pin from Bud.

Perhaps it was this whirlwind romance of his teenage daughter that prompted Mike Sloman's desire to re-enroll his daughter in school and pull her away from a relationship that was growing more serious by the day. Mike contacted both the public Burlington High School and the private academy St. Mary's Hall, both located in Burlington, New Jersey, to find a school for Laura. An undated internal memo outlines the family's wishes and St. Mary's concerns: "Laura has had a very sketchy education, but the family do not wish her to work for a diploma. I surmise she will be a very special student."

School administrators were uncertain where to place sixteen-year-old Laura, in either the eighth or ninth grade, so she was given the Kuhlmann-Anderson Tests, standardized tests still in use today, to measure aptitude and academic potential. Earning normal scores for a sixteen-year-old, Laura was admitted to the ninth grade at St. Mary's Hall in September 1941.

To Laura, education took a backseat to romance. Urging her to write as often as possible, Bud wrote: "Oh tell me lots and lots of things for when I get swell letters from you I read them over and over and over and feel good for days. Light of my eyes, heart of my heart, love of my life – for ever and ever – I Love You."

Love and Sacrifice

Laura was ever-present in Bud's mind. Fellow cadet Jack Peck brought Bud a copy of *Life* magazine with a story, "Hays Office Cracks Down on Cinema Sweater Set," pointing out one of the women bearing a resemblance to Laura. Writing to her about the magazine photographs, he told Laura, "you really are better looking."[160]

As love blossomed between Bud and Laura, war was engulfing the European continent, weighing heavily on the minds of West Point cadets. In the spring of 1941, Bud's original class knew for certain that combat was in their future as the curriculum changed to meet the rising challenges. Many academic classes were curtailed in favor of weapons training and field courses. Afternoons were filled with martial courses, such as Tactics and Techniques of Combat Arms, Tactics, Drill and Command, Branch Instruction, Riding, Field Art, and Coastal Artillery, plus flight training.

West Point *Pointer* magazine cartoon, 1942

Across the Atlantic, the British launched offensives in northern and eastern Africa against the occupying German and Italian forces. As Royal Air Force bombers pestered German cities, Germany countered with heavy bombing of British cities in the Blitzkrieg. President Roosevelt was persuaded by Churchill's declaration, "Give us the tools and we'll finish the job," proposing the Lend-Lease Act, that was passed by Congress on March 11, 1941. This provided weapons and supplies to Great Britain, while technically keeping the U.S. neutral.

With his brother and mother safely evacuated from the Philippines and safely back on U.S. soil, Bud still worried about the well-being of his father, so far away.

MacArthur was returned to active duty as a major general on July 26, 1941. The intensity of the training picked up immediately as a test for the men under the most rigorous conditions and, no doubt, it was under these conditions that Ollie fell ill. At the beginning of 1941, Ollie's job as Plans and Training Officer was taking him deeper into the

General Staff officers, 31st Regiment, Ollie center of group.

A gaunt Ollie Reed captured by a street photographer in San Francisco.

Filipino jungle for greater lengths of time. It was on one of those exercises when Ollie contracted malaria. Older officers were already being returned to the U.S. and replaced with younger men. At forty-five and suffering his second bout of malaria, Lieutenant Colonel Reed was sent home. He returned to the U.S. aboard the SS *President Coolidge*, arriving in San Francisco on November 15, 1941. This was the second-to-last trip the *Coolidge* would make from Manila.

Three weeks and one day after Ollie left Manila, the Japanese bombed Pearl Harbor on December 7. They invaded the Philippine Islands the next day. As Ollie recuperated thousands of miles away in the safety of San Francisco, his men were fighting and dying in the face of an overwhelming invasion force.

Upon his arrival in San Francisco in November 1941, gaunt but alive, Lieutenant Colonel Reed was temporarily assigned to the Third Division, headquartered at San Francisco's Presidio. By this time, Mildred had settled in Manhattan, Kansas to be close to Ted, but she came to the West Coast to meet Ollie upon his arrival in San Francisco. After a month's recuperation, Ollie received orders to report to Fort Jackson, South Carolina, with the 30th Division on December 11, 1941.

With the bombing of Pearl Harbor and the invasion of the Philippines, Mildred was happy to have her men all safe on U.S. soil, but she worried endlessly about the boys they left behind.

Again, America tried to remain neutral in the face of international

Love and Sacrifice

conflict, while bracing for war, and the inevitable came to pass with the Japanese attack on Pearl Harbor. War was declared against Japan the following day, December 8, 1941. Three days later, on December 11, war was declared against Germany. Hundreds of thousands of men and billions of dollars were being mobilized in the effort. Over the two years, from 1939 to 1941, the U.S. military budget rocketed from five hundred million to more than six billion dollars for Fiscal Year 1941. In 1942, the number rose to nearly twenty-eight billion, culminating at eighty-three billion dollars by the war's end in 1945.[161] The 1941 Mobilization Plan called for a 1.1 million-strong military force on the ground, 70,000 replacements, trained and at the ready, and 200,000 in the Air Corps, for a total force of 1.4 million, comprised of enlistees and draftees. The order of battle would be comprised of two armies, nine corps, twenty-seven infantry divisions, four armored divisions, two cavalry divisions, one cavalry brigade, and within the general staff.[162]

Since 1939, Fort Jackson had been in a constant state of expansion, handling a flood of enlistees that quickly turned into a tidal wave with passage of the 1940 Selective Service Act.[163] Men between the ages of twenty-one and thirty-five were required to register for the draft and face the possibility of induction into the military. The first wave of draftees reached Fort Jackson and other posts in October 1940. The 30th Division was a training division tasked with preparing incoming soldiers for duty. Christmas of 1941, which Ollie shared with his family and cousin Mabel in DC, was somber, with war raging on two fronts and American boys in the fray. Mildred wrote about her family coming together for that holiday:

Reed family in Washington, DC, Christmas 1941

> Pop came from Columbia South Carolina, I came from San Francisco, Bud from West Point and Ted from Manhattan, Kansas.

After early communion in the Bethlehem Chapel of the National Cathedral, we went to the restaurant at the Zoo for breakfast. There were no other customers. We pulled our table in front of the fireplace and had the full attention of the two waiters. We will always cherish the memory of the two happy hours spent there. After breakfast we toured the Zoo, never dreaming that it would someday be ultra important in our lives.

The day after Christmas, Bud took the train to Beverly, New Jersey, to spend the remainder of his holiday break with Laura on her family's farm. While his parents may have known about his visit, they were unaware of his plans.

In her diary, Laura recounted two very important days:

December 26th, 1941 Beverly Farm

Dear Diary –

My Bud came today to stay till January 1st, at which time he must be back at West Point.

I do believe I get so I love him more every visit…he is so horribly wonderful.

Bud and I have talked it over and decided that we mean a great deal to each other and we want each other a lot, and the first chance Bud gets to be with father, alone, he is going to ask him for my hand. Now that we have decided for sure when he is going to ask him, we both feel nervous about it all.

The next day she wrote:

Dear Diary –

I have had a very pleasant day with my dear Bud…he is so sweet.

Just before supper I was in the kitchen helping mother…and Bud was in the living room with father…and I knew, I don't know how but I knew, he was asking father our most important question, our whole future. I was suffering in the kitchen like a husband waiting for a first baby…I got hot then cold, and kept working…for fear I would stop…and then I don't know would have happened.

Then suddenly I calmed down as Bud walked into the room holding his fingers in the sign of victory! I believe it was the most welcome sign I have ever seen.

We all then went into supper…and after father served the soup, even before he sat down, he looked at mother and said, "Bud asked if he could marry Laura, just before supper and I told him yes!" I looked at mother and I do believe she was crying. Then Bud got up from across the table and came over to me and placed a miniature of his graduation ring on my finger and kissed me. I didn't know whether to laugh or cry…I guess I did a little of each at the same time.

Laura and Bud, Army-Navy football game, December 29, 1941.

It is the most beautiful engagement ring in the world and I do adore it.

Its funny that I should sit here and write this to you dear diary, for I know I shall never forget a minute of this day.

…I just know we both shall be very happy together.

★★★★

Bud had not told his parents in advance of asking for Laura's hand in marriage, and did not tell them until March of their plans to marry the day after Bud's graduation. Mildred, in recounting her son's engagement, did not reveal her own feelings, which were not overly positive, but told a slightly different story than Laura's diary.

Buddy asked her father in December 1941. Laura turned 17 on the 30th. Mike said they had his permission if they'd wait two years.

When Buddy's graduation from West Point approached, they decided they didn't want to wait. Laura asked her mother, Katherine (who was a step-mother but she didn't know it.)

Katherine had always been jealous of Mike's deep affection for Laura. She told Laura to leave everything to her. She made plans, invitations and had everything ready before she told Mike.

As a Firstie (a cadet in his final year at the academy), he was now gleefully joining in the nonsense that had shocked him during his earliest days at the academy, and he was paying a price. "Throwing glass in Mess Hall, noon meal" - five demerits and ten confinements to quarters. Even in his fifth year, promoted to cadet sergeant, Bud still racked up the demerits for routine infractions: boots improperly shined, clean laundry not put away, laundry bag concealed in barrack policeman's locker, room in disorder, absence card not properly marked, and late breakfast formation, among many others.

One instance occurred on Sunday, February 22, 1942 when the Cadet Corps assembled at 8:50 a.m. for church service. The combined choir and corps sang "Holy Thou Art" and "Alma Mater," finishing with the "Star Spangled Banner." As they marched from the chapel and broke ranks, Bud was apprehended as he tried to sneak away with Laura, who had attended the service as part of the general congregation. Bud was assessed a five-and-five (five disciplinary points plus five hours walking tour) for "escorting during Chapel service."

Writing of his love for Laura in 1942 ahead of the wedding, Bud penned:

> If I were an artist I would paint a great picture; if a poet I would compose a deathless song of love; if a musician a melody to stir your heart. But I am none of these. I am just an ordinary, simple man, and so there is no way I can express the inspiration you are to me except to tell you that – I love you.
>
> There is no way for me to adequately express, explain or tell you of the passion that lives in my heart. I can only show you by my every word and every action through all the long years of my life that you are the one woman in the world for me – that for your love I would gladly die.

Love and Sacrifice

The war was accelerating everything. Graduation and "June Week" for the class of 1942 were pushed up two weeks to May 29, and their first leave as second lieutenants was cut from three months to two weeks.

One event that had not changed was Branch Night when, immediately following the conclusion of the academic year for the first class, the "order of merit," their class ranking, was posted. This determined which branch of the army the new second lieutenants would be assigned. The highest ranked had the option to join the Corps of Engineers, an honor. Given his scores, Bud already knew he was destined for the infantry.

In advance of the wedding, Mildred wrote to Laura, whom she had not yet met, complimenting her on the choice of a small wedding and wishing her happiness: "It won't be long now, will it! The days do slip right along. I can imagine how excited you are over the 'big week' coming up, and I do hope and pray that you will be just as happy as your fondest dreams."

While it opened with pleasantries, Mildred's letter conveyed underlying tension and disappointment; her beloved son was marrying someone his mother had never met and it was all happening in such a hurry. Laura was not Nancy, whom Mildred loved as a daughter and had looked forward to welcoming into the family. With such sentiments thinly veiled between the lines, Mildred's feelings about some dishware purchased as a gift for the couple were not so subtle: "Bud tells me you took back the English dishes. Was it because you couldn't finish out the set? It was certainly an attractive shade – and so unusual. Bud says you got some electrical equipment. That is fine."

Mildred included some wartime prognostications, maybe thinking she was adding a positive note in the letter, but no doubt her words sent shivers up and down the spine of the teenage war-bride-to-be: "I heard the other day that Wall Street was betting the war would be over in October! I fear they are optimistic – if they only meant October 1942 – (maybe they meant 45) but it wouldn't make us mad if it ended tomorrow would it!"

Topping off the letter, Mildred threw a wet blanket on Laura's wedding plans, talking about how the newly imposed gas rationing might put a damper on the number of people attending, and that others might be busy with other events, adding, "lots of people will be disappointed – but it is a small thing, really, compared to what our boys in the Far East are doing for us."

Mildred, Bud, and Laura before West Point graduation.

Mildred had a tough time hiding her feelings. A photograph of a proud and happy Lieutenant Reed, standing between his mother and his nervous fiancée on graduation day, shows Mildred with a rather dour expression. Laura was too blissful to notice, writing in her diary on May 26 about Bud's parents and brother picking her up that day for the drive to West Point, "I was plenty scared but pretty soon relaxed and I like them all." Mildred captioned a picture of Laura in her photo album as Bud's "OAO" - "One and Only."

The next day started with the Baccalaureate in the cadet chapel and then a family dinner at the Thayer Hotel, followed by a dance.

"Father Reed got his new orders right after supper and it surprised me so that I cried," Laura wrote in her diary, "and he saw me and told me he thought I was a swell girl and that Bud was lucky. I think he is wonderful. I hope Bud is like him (at that age)."

It was difficult for Ollie to get leave to attend Bud's graduation and wedding, but the army knew exactly where to find him. His new orders were to report to the 78th Division at Camp Butner, just outside of Durham, North Carolina, as the G-4 officer in charge of supplies.

Graduation festivities continued with the superintendent's reception the following afternoon and the big Graduation Hop, exclusively for graduating cadets, featuring Kay Kyser, a very popular bandleader and radio personality known as the Ol' Professor of Swing, with his Kollege of Musical Knowledge.

The commencement was held on Friday May 29, 1942 inside what is today the Gillis Field House, named in honor of 1941 graduate Major

OUR SOLDIERS WILL LAND IN FRANCE, MARSHALL TELLS WEST POINT CLASS

Headline from the next day's New York *Times* that Mildred saved in her scrapbook.

William G. Gillis. U.S. Army Chief of Staff General George C. Marshall gave the commencement address. Marshall made headlines with his pronouncement that American troops would soon be landing in France: "We must be prepared to fight anywhere, and with a minimum of delay. The possibilities were not overdrawn, for today we find American soldiers throughout the Pacific, in Burma, China, and India. Recently they struck at Tokyo. They have wintered in Greenland and Iceland. They are landing in Northern Ireland and England, and they will land in France."

1942 West Point graduation (USMA)

The graduating cadets rose to give General Marshall a rousing ovation when he said, "We are determined that before the sun sets on this terrible struggle our flag will be recognized throughout the world as a symbol of freedom on the one hand and of overwhelming force on the other."

As Marshall neared the conclusion of his speech, Mildred could not help but shed tears for many families she knew so well, especially when the general said:

> The calm and the fortitude with which they accept the vicissitudes that are inevitable in a struggle that goes to the four corners of the earth are very reassuring. And our greatest reassurance comes from the courage and fortitude of the wives and parents of those who fought to the last ditch in the Philippines.
>
> I do not know of anything which has impressed me so much with the present implacable state of mind of the American people as the letter I received from the wives and mothers of those men

in the Philippines who went down in the struggle, either as casualties or prisoners. Their heroic messages of fortitude and resolution are an indication of the fact that this struggle will be carried to a conclusion that will be decisive and final.[164]

Mildred had been thinking back to Bud's first year at the academy as a member of the class of 1941 and how different things could have been if he had not failed French:

> We were very disappointed, but remembered our verse, Romans 8:28, 'All things work together for good...' As always, it proved true. If Bud had graduated in 1941, he would have come to the Philippines where we were. Half a dozen of his friends did. In a few months they were casualties.

Proud father and his new lieutenant, West Point graduation day, 1942.

Years later, Mildred proudly remembered one particular moment in the graduation exercise:

> As the graduates went down the receiving line for congratulations from the General, he stopped Buddy, asked if his father was Ollie Reed who had commanded the machine gun company of the 29th Infantry at Fort Benning; sent good wishes to his father and congratulated Buddy for following in his father's footsteps. This was before hundreds of our army contemporaries! What a memorable moment! Bless General Marshall!

The new second lieutenants streamed outside to congregate and congratulate one another. For four years, some more, they had lived and struggled together, and now were being cast into their first real assignments - on the battlefield. Of the 502 young Beasts who entered West Point in the class of 1942, 374 made it to graduation. This was the first West Point class graduating into World War II and the last full-term class for the remainder of the war.

War could wait, though, as the freshly minted second lieutenants

The graduate with the ones he loves best.

had other things on their minds, for many it was getting married. The cadet chapel had been booked since March for graduation day weddings that would take place every fifteen minutes, lasting until 10:30 that night, with more the next day and in other locations. Jay Hewitt, a native of Grand Forks, North Dakota, was the first to wed, at 12:30 p.m. His marriage to Aulene C. Cunningham, West Point's hospital dietitian, was featured in *Life* magazine. On that day, 114 weddings took place. Companies I and L each had thirteen walk to the altar, while ten from Bud's Company F stepped through the chapel door under the canopy of crossed sabers.

Bud and Laura opted for a small family wedding at the Sloman's farm in Beverly, New Jersey. After graduation, they hopped into Bud's blue 1941 Plymouth coupe, given to him as a graduation gift by his parents.

Bud christened the car, "Bluebird of Happiness," and in it, the happy couple drove south to New Jersey with Bud's parents and brother close behind for the wedding the next day.

Laura's January 22, 1942 diary entry.

"I shall never forget the night before the wedding the boys and Pop and I knelt by our bed and Bud's fervent prayer was that we all might be together again after the war - if it was God's will," recalled Mildred.

At 11:00 a.m. the next day, Laura, tightly clutching a handkerchief in her right hand and her father's hand in her left, walked down the stairs to the living

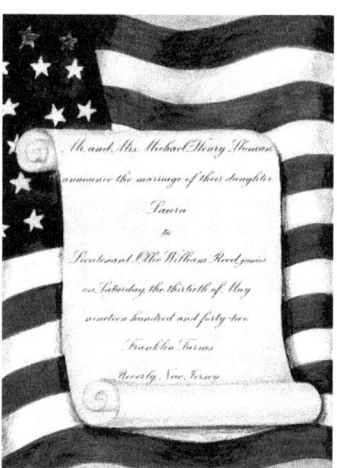

Laura and Bud's wedding announcement.

Catherine Sloman, Lenora Velie, Laura, Bud, and Ted. Below, boarding the Bluebird of Happiness.

room where Bud, sharply dressed in his new U.S. Army uniform with single gold bars on his shoulders, stood next to the Reverend Roy Williams, pastor of the Rancocas Methodist Church, with his best man, brother Ted, and Lenora Velie, Laura's best friend and maid of honor.

Everyone was beaming as the bride walked toward her groom across the room. In spite of darkness enveloping the globe, all in attendance shared in the happiness of this moment.

After a wedding luncheon, the bride and groom climbed into the Bluebird for their Florida honeymoon. It would be a quick trip. Bud and sixty other West Point graduates had only two weeks before they were to report to the Infantry School at Fort Benning on June 14, 1942.

Newlyweds with their parents

Fathers, groom, Jim Sloman, and Louise

Ted, Laura, Bud, and Lenora Velie

Laura and Catherine; Bud and Laura; Margaret and daughter Louise; Aunt Mabel Reed and McCullys.

CHAPTER SEVENTEEN

Starting a New Life

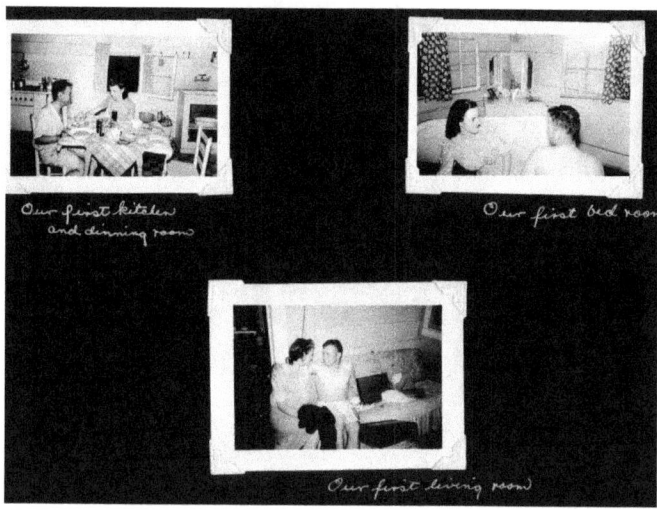

ON THEIR WEDDING DAY, the newlyweds drove only as far as a cabin outside of York, Pennsylvania, on their wedding day, settling in for their first night together. Laura recounted the night in her diary:

> May 31, 1942 York, Pa.
>
> Mileage 12,825
>
> Cloudy & cool
>
> Dear Diary,
>
> Bud and I got out of our nice cabin about 8:45. I can't believe we are married – it was heavenly to sleep with him – all tho I was sick.

They drove along Virginia's Skyline Drive to Swift Run, Virginia. Skyline Drive runs the north-south length of the Shenandoah National Park along the peaks of the Blue Ridge Mountains. Swift Run Gap was, and is, one of only four entrance and exit points along the 105-mile road. The closest town to Swift Run Gap is Elkton, where Bud and Laura spent the night at the newly renovated Gables Hotel.

With so little time for a honeymoon, on the second day they had to cover a greater distance than they did on the first. Making up time, they sped along U.S. Route 11, crossing the Tennessee line. Laura noted the memorable day:

> 6/1 to Kingsport, TN – big clean cabin for $1.50 and a huge supper for 90 cents; made love for the first time.

She was comfortably adjusting to married life, telling her a diary a little secret as the newlyweds arrived at their destination of Panama City, Florida, on Sunday June 7.

> Diary, don't you let on I told you but we are having a wonderful life and honeymoon. We make love twice a day.

After a week on the beach, Bud and Laura drove to Columbus, Georgia, where they set up house in a small two-story bungalow on Hamilton Road, which they rented for twenty dollars. Bud had to report to Fort Benning the next day to begin ninety days of training at the Infantry School.

The young couple was so thrilled with being married and setting up their first home, they set up a camera in each room to photograph themselves in each situation.

On August 26, Laura visited an army doctor who told her that she might be pregnant.

"My perfect husband is acting like he will be a father tomorrow," she wrote in her diary.

Bud graduated from the 36th class of the Infantry School on September 8. The band played the "Beer Barrel Polka" as the graduates received their certificates. The same day, Laura received the sad news that she was not pregnant. They immediately left Columbus for Bud's assignment at Camp White, Oregon. Ted, who was visiting, hitched a ride as far as Kansas. It was a long cross-country trip, made all the longer by the mandatory thirty-five mile per hour speed limit imposed due to wartime gas rationing and the rubber shortage.

Bud was assigned to the 91st Infantry Division, known as the Fir Tree Division in the First World War, a newly reactivated division headquartered at Camp White near Medford, Oregon. The division's battle cry, also from the previous war, was "Powder River! Let 'er Buck!" Camp White, named in honor of Oregon National Guard Major General George White, who had just recently passed away, was one of nine training camps ordered by the War Department to be hurriedly constructed in 1941. Construction began on February 25, 1942, and was still underway when Second Lieutenant Reed arrived.

Major General Charles H. Gerhardt was the commanding general of the 91st Division. He had arrived at Camp White on July 8 and by July 19 most of the division officers had reported for duty. The 91st was officially reactivated on August 15, 1942. Bud's father knew General Gerhardt from their days together at the Command and General Staff School, but that would curry no favor for Lieutenant Reed. Gerhardt was tough and demanding, perhaps best known for ordering the 91st on a ninety-one-mile hike in September 1942 through the arid Oregon hills, over the rough roads, and trails of the Cascade Mountains, in under thirty hours. Previous marches along this route had taken six days.

The 91st Division marching through the hills around Medford, Oregon. (USHEC)

Gerhardt's slogan was "March, Shoot and Obey!" and he demanded "results, not alibis." Some of the men referred to Gerhardt's program as the Alcatraz of training camps, but the general insisted that, with his training, more men would come back from the warfront alive.

In October and November, recruits and draftees streamed into Camp White, totaling twelve thousand new and untrained men. As men arrived at the train station, they were herded into groups as per their assignment. Training was a rude awakening; the first order of business was to learn how to properly march. And, march they did, from the train station to Camp White and their company barracks, through an autumnal Pacific Northwest pouring rain. Fall turned to winter, as the rain, mud, and snow continued relentlessly.

Most of the men had no previous military training and needed to be drilled in the basics of close order drills, standing at attention with their stomachs pulled in, chests puffed out, and heads held high – all too familiar to West Point graduate Reed. Thumbs were pressed on pants seams while heels were pressed together, toes pointed out at a forty-five-degree angle. Few escaped verbal haranguing by an officer over the slightest flaw. For Buddy, the grueling lessons of his Plebe year came back in a rush. He knew these drills and also how to teach them to others, but the "ninety-day wonders" quickly produced by Officer Candidate Schools knew the basics of this basic training but had difficulties conveying it to others. A cadre of rough and tumble sergeants from Fort Bliss, Texas were brought in to help with the training tasks. These were hill country sergeants who knew the army from the ground up – Texas style.

The learning curve was steep and rapid-fire over the thirty-nine-week basic training that began on November 15th and concluded on February 15th, followed by another thirteen weeks devoted to tactics.[165]

On Christmas Eve 1942, Laura wrote in her diary: "Bud had to stay at camp tonight but I went out to see him until 10:00. When I got home I placed Momsie's gifts under our tree."

Bud was able to spend Christmas with Laura and made the best of whatever time he had in the spring. As a second lieutenant, Bud earned just $252 a month as a new second lieutenant, barely enough to pay the rent and utilities on a farmhouse near Camp White. To supplement this, Bud and Laura bred cocker spaniels birthed by their beloved black cocker, Duchess Nebbie.

Bud's separation from Laura for weeks and months on end was

miserable, made all the more painful by Laura's announcement in the spring of 1943 that she was pregnant – this time for certain. In what Mildred described as her "reddest letter Mother's Day" in 1943, she received the call from Buddy with the news that she was going to be a grandmother.

Pregnant Laura with Duchess Nebbie's puppies.

Bud spent much of that summer on maneuvers, when the 91st joined the 96th and 104th Divisions for maneuvers covering nearly ten thousand square miles of hot, dusty desert, and cold mountainous terrain. In November, the division moved from Bend, Oregon to Camp Adair outside of Albany, Oregon. At the same time, Bud was anxiously awaiting the birth of his child.

In November 1943, Mildred took the train to Oregon from North Carolina where Ollie was stationed with the 78th Division at Camp Butner, so she could be with Laura as the birth day approached.

> Laura and Buddy lived in a farm house near Camp White, Oregon. The baby was due in December. It was lonely out there, so I went to be with them. Arrived in time for Thanksgiving. Buddy and I cooked rabbit for Thanksgiving dinner. It cost $1.35. We used to buy them for a quarter. Laura had never eaten rabbit before and wasn't enthused, but she was game and accepted it as another one of the crazy ideas these Reeds hatched. The baby was a month later than expected. Laura was pretty miserable those last weeks. Buddy was feeling sympathetic. I heard him say, "Are you sure you want to go through with this?"
>
> Pains finally started early on Jan. 18, 1944. I was scurrying around getting my writing materials together, since I knew we'd be in Corvallis, where the hospital was, several days. Bud came from the kitchen eating a breakfast roll. Laura was disgusted with us. "You two!" she exclaimed. "Here I lie, practically dying and Bud would eat and Mildred write letters." We got her to the hospital safely. I remember Duchess Nebbie pacing the ground beneath her window, whining, wishing she could comfort Laura. When Bud

and I went out for lunch, he bought a Rook deck. When she was in the throes of labor pains, he said, "Come on now. Get your mind on something else. Let's have a game of Rook." I am afraid he was remembering the expression he often heard in days of his youth from his mother's lips, "It's all in your head." The idea didn't go over well with Laura.

Toward the end of the afternoon she turned to me, crying, "You didn't tell me it would be this awful." I said, "I know I didn't. You can't believe it now, but you will forget it." The nurse brought the baby - ten pounds, four ounces - to the window of the waiting room. Never was there such a wonderful child! How Buddy beamed!

JAN. 28 1944 PROUD PARENTS

When we got home from the hospital, Buddy took over night duty.

I slept on a couch in the living room and feigned sleep when he'd bring the baby out to change him in front of the fireplace. I was amazed how Buddy took to fatherhood, never having been exposed to babies. He'd talk to the baby, gently reassuring him.

"Now calm down, Son. Everything's going to be all right. There. There. Daddy's right here and he loves you."

A cherished picture in my mind is when Bud came home from work, he'd sit before the fire holding the baby reading aloud to him from Peter Rabbit. He was trying to crowd a lifetime into the few weeks he knew that he'd have with him.

On January 28, 1944, only ten days after Ollie W. Reed, III was born, the War Department issued the formal movement orders, setting a readiness date of March the first. Mildred wrote:

When the baby was three weeks old, I left Oregon for Blackstone, Virginia, near Camp Pickett, where Pop would be coming from maneuvers. Laura and Buddy took me to the railroad station in Portland. Buddy had it figured mathematically that as a lieutenant from West Point he would probably not come back. He told me he wanted me to encourage Laura to marry and have a normal happy life. I was weeping after he left. I knew I'd never see him again. He came back into the station for one last kiss. He said "Now Mother, you've got to trust the Lord to take care of me." My child telling me that, when I am the one who should have been reassuring him!

By March 8, everything was ready for deployment. Bud wrote to Laura as he prepared to board a train heading to points unknown. Everyone in the 91st believed they would take a short ride to a ship waiting to carry them across the Pacific Ocean to the fight against Japan, but no one, except for the division commanders, knew where they would land for certain. Laura had already left, heading east to visit her family in Philadelphia. There was nowhere else for her to go in the west, so she went home.

3/28 Sunday night

Camp Adair, OR to Phila

My Sweetheart,

I did not write to you yesterday nor the day before. The regimental post office was shut down so no letters could be mailed. Everyone complained about this so today the Post Office was reopened but it will close for good tomorrow night.

I truly believe this will be the last letter you will get until my train gets to wherever it's going. I will write every day on the train, and have lots of letters to send you when I get there. Please be patient with me, dear. I know my letters come very irregularly but everything is in turmoil here. I will be glad to get on the train for there will be more peace there.

I cannot call you again. The outside phone is off limits now. The night I called was the last time it could be used by my regiment.

You have been gone a week now. Tonight at 6:27 p.m. I stopped and thought of last Sunday. My heart aches at the thought. My dearest darling, lover I adore you so. I still don't see how we parted.

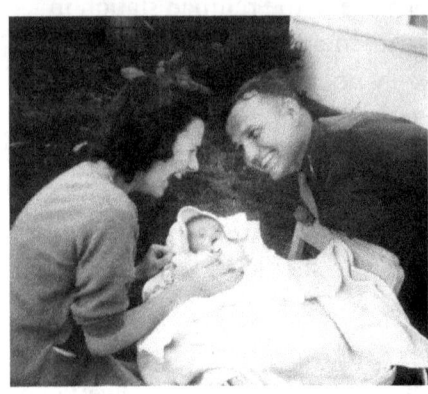

My arms clung to you tightly and wouldn't let go. That last minute came so quickly – so dreadfully quickly. My eyes fill with tears now as I write of those dreadful last seconds that went so fast. My life is empty without you, Laura - it always will be empty while you are gone.

I love you. I love you. I love you. There is one phrase from a song that I now know the full meaning of – "You are always in my heart."

You are, my lover, you truly are always, always in my heart. You would be very surprised to see me, dear, for I am crying. I never, never used to cry for any reason but now it is very easy for me to do so. Almost any song we ever sang together fills my eyes with tears – "I'll Get By," "As Time Goes By," "You and I Know the Reason Why," "Deep Purple" – and many more.

I am glad you have the feeling that I will come back to you. I believe that too. Honest, I do.

Yes, definitely want you to raise Ollie as you see fit. He is our baby and no one else's. You re a good mother and you have a lot of brains and you know how we want Ollie. We both agree on everything important so you can always act with my full approval and backing.

My fondest dream now is a home for you and Ollie and me – a real, honest-to-goodness place of our own that we can live together all our life. That is worth much more than traveling around from place to place in the army. I love you and want to be with you and loving you always.

You are a perfect wife for me. You are beautiful – you have lots of brains – you dress well and talk well – you run our home so smoothly – you keep a lovely house and you are always a grand cook. You cheer me up when I am blue – you talk to me and make me feel you really care. In fact, when you talk to me about everything you make me feel we are just one person. And then, at last, when the day is done you are a grand and passionate lover...

You are perfect for me, my lover – I adore you in every way.

"Goodnight, sweetheart...sleep will banish sorrow." You are the only one for me.

Your

Bud

You are always in my heart.

Rumors floated among the rail cars, changing with every leg of the journey – somebody heard from somebody else who overheard...

In truth, only a few officers knew their destination and none would talk outside the circle of command.

It didn't take long for the men of the 91st Division to realize they would not be sailing across the Pacific as they headed east over the Cascade Mountains to the welcomed sight of a sign reading, "Welcome to Boise." They were relieved to not be thrown into the Pacific fight. To the east was Europe, but that was as far as their imaginations could wander.

While Bud and the 91st were riding the rails, Colonel Reed and the 78th Division moved north from Camp Butner in North Carolina to Camp Pickett in Virginia for maneuvers. Mildred joined Ollie three weeks after the birth of Ollie III. Laura and the baby drove to Blackstone, Virginia, near Camp Pickett, from her parents' new home in Philadelphia. Bud secured leave from his division's destination at Camp Patrick Henry in Virginia to join them in Blackstone.

Ollie, Mildred, Bud, Laura, and Ollie III spent their last evening together in a heated game of Monopoly, one of Bud's favorite activities.

Major General William G. Livesay, center, commander of the 91st Division, with members of his headquarters company.

The first contingent of the 363rd Infantry Regiment of the 91st Division sailed from the Hampton Roads Port of Embarkation in Virginia on April 10, 1944. The departure date for Bud's Second Battalion was delayed until April 21st due to a "lack of shipping space," giving the young lovers some extra time together. Over the eleven days, Bud took leave whenever possible from his posting at Camp Patrick Henry, located in a low-lying area near the Chesapeake Bay, nicknamed, "Swamp Patrick Henry." Damp and slippery boards provided walkways over the mud leading to barracks smelling of antiseptic that were heated by pot-bellied stoves. The men caught their first sight of POWs who freely roamed the grounds of the camp.

In an April 14 letter, written after spending time that day with Laura, Bud inked on the back of the envelope: "*Do not open – except in case of death! April 18, 1944 Promise!*"

There is no way of knowing whether Laura abided by Bud's wish, but inside the envelope, Bud shared his fears of not returning – words that would have deeply stung Laura.

> We will be separated now for a time. How long we do not know, but it will be longer than any time we've been separated before. But it will not be unbearable for we know that in time we will be together again. Seconds grow into minutes and minutes into hours.
>
> The days and weeks and months go by steadily and surely, and we will soon welcome each passing day with the knowledge it is one less day between us. Yes, the pain of separation is bearable. The

only unbearable pain is one that never ends.

That leads us to the next thing I want to talk to you about – what happens if I don't come back. Death does not hurt the one who dies; death only hurts the lover. It would not hurt me to die; it would hurt you. I know how I would feel if you died. In fact, I remember how I did feel the day I held your hand as you went into the valley of death for my sake. So you must be prepared for a great shock and a great pain if God wills it. But you must also be prepared and make up your mind now that you will not give up – that you will see it through with all the guts you have because I want you to and because Ollie needs you. The first day will be the hardest day. The first month will be the hardest month, the first year will be the hardest year. Time will keep moving by steadily and mercifully; it always does.

Mildred remembered the last moments Buddy spent with his son:

"The morning he left, he took little Ollie out of his cradle, kissed him and said 'Grow up to be a good honest man.' I held the baby up as he drove out of sight so his last view of home would be of his precious son."

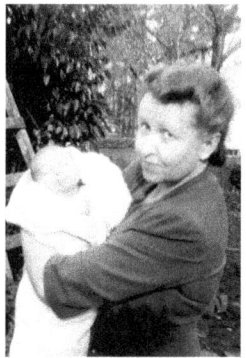

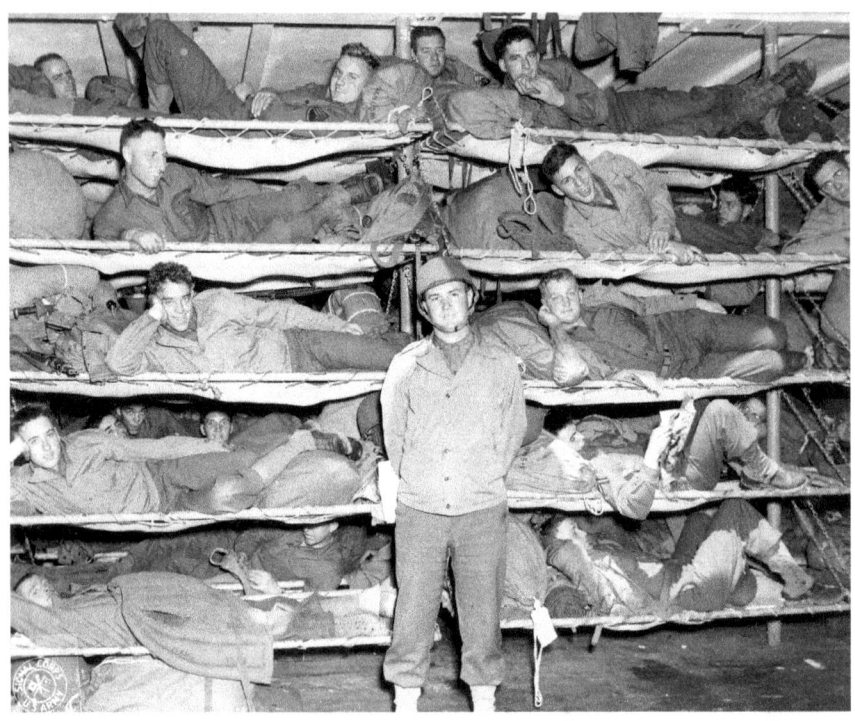

1st Lt. Chetlain Sigmen and Company K, 3rd Battalion, 363rd Regiment, 91st Division aboard HR-103 before sailing overseas. (The Mariners' Museum, Newport News, VA)

CHAPTER EIGHTEEN

Departures

Atlantic convoy (Library of Congress)

BUD'S CONVOY, DESIGNATED UGS-40, started with eighty merchant ships, and was escorted by four destroyers, eight escort destroyers, one anti-aircraft cruiser, a patrol boat, two minesweepers, a Coast Guard cutter, and one tugboat. By the time they crossed the Atlantic, the convoy had grown to 101 merchant ships plus escorts. Bud was one of 500 men crammed into the hull of the SS *William B. Giles*, already packed full with supplies and equipment. The merchant ships, known as Liberty Ships, had originally been built as cargo vessels but were quickly converted to troop carriers to meet the growing personnel transport needs. The galley and mess facilities were inadequate to serve the hundreds of men on board, as were the sanitary installations. Food storage was under capacity and fresh water in short supply. Men had little to do other than sun themselves on the deck, read books, play cards, and hold sporting events, such as boxing and wrestling matches, while spending much time writing letters home.

"I spend a lot of time on deck in the sunlight and fresh air. The sea is a beautiful deep blue, and the wind is very strong. Whenever I look out at the convoy I think of you and wish you could see these things too," Bud wrote to Laura. "Boxing matches and wrestling matches and skits or little plays. Also music and songs. We have some good banjo and ukulele players on board as well as harmonica players. The officers all chip in and buy prizes for the winners of the boxing and wrestling matches."

Bud didn't smoke or drink alcohol, and while he was an avid game-player, he enjoyed board games, such as Monopoly and chess, not gambling. He was thrilled to play what he called "real poker – no cards wild and only 5 card stud or 5 card draw." Starting with forty dollars, he played until four a.m. and excitedly wrote to Laura that he was "as hot as a pistol," winning $140.

"I will put $100 away and not touch it again for the rest of the trip. I will try."

Then he lost $20 and then another $20.

"I will not play any more poker so I will have $100 to send you at the end of the trip."

Bud was true to his word and as soon as he landed he mailed the $100 "anniversary present" to Laura.

One game Bud did excel at was chess. He started teaching the men on board to play the game, and pretty soon there were ten games underway. He organized a tournament in which the winner had to play the West Point lieutenant, since he was the best. Bud won.

As the converted cargo ship had no security staff or guards, the army had to supply their own. Bud was named adjutant, placing him in charge of guard assignments. The absence of crew to adequately serve and police the passengers aboard the hastily converted ships constantly posed problems for the Coast Guard, Navy and Army throughout the fleet. The cargo ships, manned by Merchant Marines, were staffed at their cargo-hauling levels, so the men working the ships were adept at moving crates and equipment, but not suited to transporting people, or to keeping order among bored and sometimes agitated soldiers heading to war. So, the army had to police their own aboard the Liberty ships.

In the midst of the wide Atlantic, the men had no idea what their destination would be. Back in Oregon, everyone was certain they'd be joining the fight in the Pacific. Crossing the country and now at sea, the bets ranged widely – England? Naples? Palermo? Greece? Their original orders had called for them landing at Augusta, Sicily, but that had changed while they were at sea. After nearly three weeks on the Atlantic, the convoy passed through the Strait of Gibraltar on May 9th, and entered the Mediterranean Sea. Six of the ships carrying the 363rd Regiment, including the *Giles*, began moving south, away from the Spanish coast and to a landing at the port of Oran, Algeria, while the rest of the convoy continued east.

As the main part of the convoy sailed past Algiers, suddenly, out of the setting sun, a German aerial attack descended upon the ships. Although the Allies controlled North Africa and Sicily since Axis forces surrendered the territories in 1943, a convoy the size of UGS-40 did not pass unnoticed by the Germans. Sixty-two Junkers 88s, Dornier 217s, Heinkel 111s, and an FW Kondor reconnaissance plane were launched from southern France to attack the convoy. The British RAF scrambled fighter planes from their North African bases to help fend off the attack. The battle lasted forty minutes, during which the Germans launched ninety-two torpedoes and numerous bombs from their aircraft. None hit their target. Seventeen German planes were shot down and the British lost two aircraft in the fight.[166]

The men of the 363rd Regiment, climbing down sisal rope ladders alongside the *Giles*, fortunately were not among those attacked. At that moment, they were too busy hanging on for dear life, weighed down by their pack and rifle, at what felt like a skyscraper's height above the awaiting DUKW, or "ducks," essentially, amphibious trucks, bobbing in the waves to take them ashore. Well, not quite ashore – rather, the men were dispatched from the landing craft into the warm Mediterranean waters to wade ashore with their rifles held high overhead. Once on dry land, the men were intrigued with the sights, sounds, and smells of this new land, including the onrush of Algerian boys wielding boxes and brushes, barking "shine, Joe?"

From offshore, the buildings of Oran shone like white beacons. Close up, the soldiers were disappointed by the dinginess and extremely poor living conditions.

Almost as soon as the Second Battalion hit the ground, Bud was transferred from Company G to Company F, led by Captain Eugene E.

Crowden, the officer in charge of security on the *Giles*.

The bulk of the 91st Division made the best of a desert encampment in Port aux Poules on the outskirts of Oran. Sea air rusted gun barrels and fine sand from the Sahara made its way into everything. In spite of round-the-clock sentries guarding the army encampment, local scavengers still managed to sneak in and steal practically anything that wasn't nailed down. Added to the joys of their desert outpost was a plague of locusts.[167]

Soon, the 91st Division would start amphibious training for a full-scale mock invasion of Arzew Beach, about twenty-five miles to the east. The training exercises would be quite aggressive, and in open daylight, easily seen by German surveillance headquartered across the Mediterranean in southern France. Unbeknownst to the 91st, this was a goal of their training – a diversion to convince the Germans that the coming invasion of France would be aimed at the Mediterranean coast, not the Normandy shoreline. The amphibious mock invasion of Arzew was part of a subterfuge called Operation Vendetta, pointed directly at the port of Marseilles. Just as the American First Army stood poised at Dover, England, within easy striking distance of the port of Calais, as part of Operation Bodyguard, Vendetta was designed to lead the Germans to believe the invasion plans pointed directly at southern France.

★★★★

Ollie was promoted to full colonel in November 1942 and given command of the 309th Infantry Regiment of the 78th Division. The rank of colonel fit Ollie to a T. Administratively, he knew how to delegate by instilling confidence into every rank of officer, NCO, enlisted man, and draftee under his command of their responsibilities and role. He led by example, upholding the lessons he'd learned in the army since 1916, and he knew how to impart those lessons to others. Colonel Reed was a beloved commander, a father figure to many. His full, jolly face exuded friendliness, even when dealing with disciplinary matters. He was regarded by all ranks as firm but fair. Fellow officers respected him, and those under his command looked up to him. To the young women marrying members of the 309th shipping off to war, Colonel Reed was often their stand-in father, giving away the bride.

Private Luther Davis of South Carolina, a talented artist in the 309th, was commissioned by regimental officers to paint a portrait of Colonel Reed from photographs the officers had given him. The presentation of the painting by Private Davis to Colonel Reed was featured in the local newspaper. Today, Mr. Davis remembers the awe and admiration he had for Colonel Reed. He recalled the kindness and respect shown to a mere private by the colonel and his wife that day. The portrait has remained a centerpiece in the Reed family ever since.[168]

The 309th was a replacement regiment. When combat units needed new men, regiments such as the 309th were called upon to supply fresh personnel. Luther Davis recalled being away on leave and returning to find most of the men had been shipped out.

Once again, Ollie was training men for battle but was never able to lead them into battle, and he let his frustrations about being left stateside known to fellow officers and to the Chief of Infantry, Major General Courtney H. Hodges. Lack of command assignments was a factor in promotion stagnation. As it was noted in his West Point memorial, Colonel John H. Van Vliet, a colleague of Ollie's through several stages of their careers, found his promotion potential held down by frequent assignments as student and teacher, rather than in active command.[169]

Colonel Reed's request for an overseas assignment was granted. Before he left the 309th, there was a parade in his honor. It was a poignant moment when he reviewed his regiment for the last time. He proudly held his salute while national, division, and regimental colors passed by. It was an emotional experience for Ollie. Sergeant Vincent Garramore told Mildred, "Even strong soldiers like Colonel Reed cannot hide their feelings in times like those. His voice was husky when he told his men good-bye after the parade."

The regiment also held a farewell dance in his honor. Mildred fell ill, so Ollie took Laura as his date. On this occasion, in the modest house on Amelia Street in Blackstone house they were sharing, Mildred remembered her husband committing a grand faux pas that he always feared he would:

> After Buddy married Laura, Pop had said he was always afraid he'd call her Nancy. This was the time. They were dressing. He came out of the bedroom. "Are you about ready, Nancy?" Laura's bedroom door was open. I was on the living room couch, but she couldn't hear anything he said. I made wild motions, but Pop wasn't in tune. When he came out, he said, "It's time to go. Are you ready, Nancy?" Laura flounced out of her room, "Yes, I'm ready, but I'll have you know I'm not Nancy."

In order to attend his father's farewell in May 1944, Ted was personally excused by KSU president, Milton Eisenhower, to take a leave from his studies to join his mother and father in Virginia. Ted and Mildred drove Ollie from Blackstone to Washington where he caught his flight to England and the war.

On May 19, 1944, Ollie boarded a plane for England and duty with XII Corps in the town of Bewdley, awaiting the invasion of the European mainland

As he was leaving, Ollie took Mildred's chin in his hand and said, "Now smile. I don't want my last look at you to be tearful."

Love and Sacrifice   235

Ted returned to Kansas and Mildred drove back to Blackstone where she joined Laura and the baby. She later recalled a fond incident after returning:

> Not long after Pop had left for the war zone, Laura and I were coming home from the Camp Pickett commissary. We passed some 309th soldiers walking to Blackstone, and we picked them up. One of them said, "Mrs. Reed, do you want to know what we soldiers called the Colonel?" I said, "I am not sure I do. I know what most soldiers call their commanding officer." He said, "We call him Uncle Ollie." How perceptive of the soldier boys to see through the military exterior to the tender heart of my Ollie!

Scenes from their time with the 309th Regiment, 78th Division from Mildred's scrapbook.

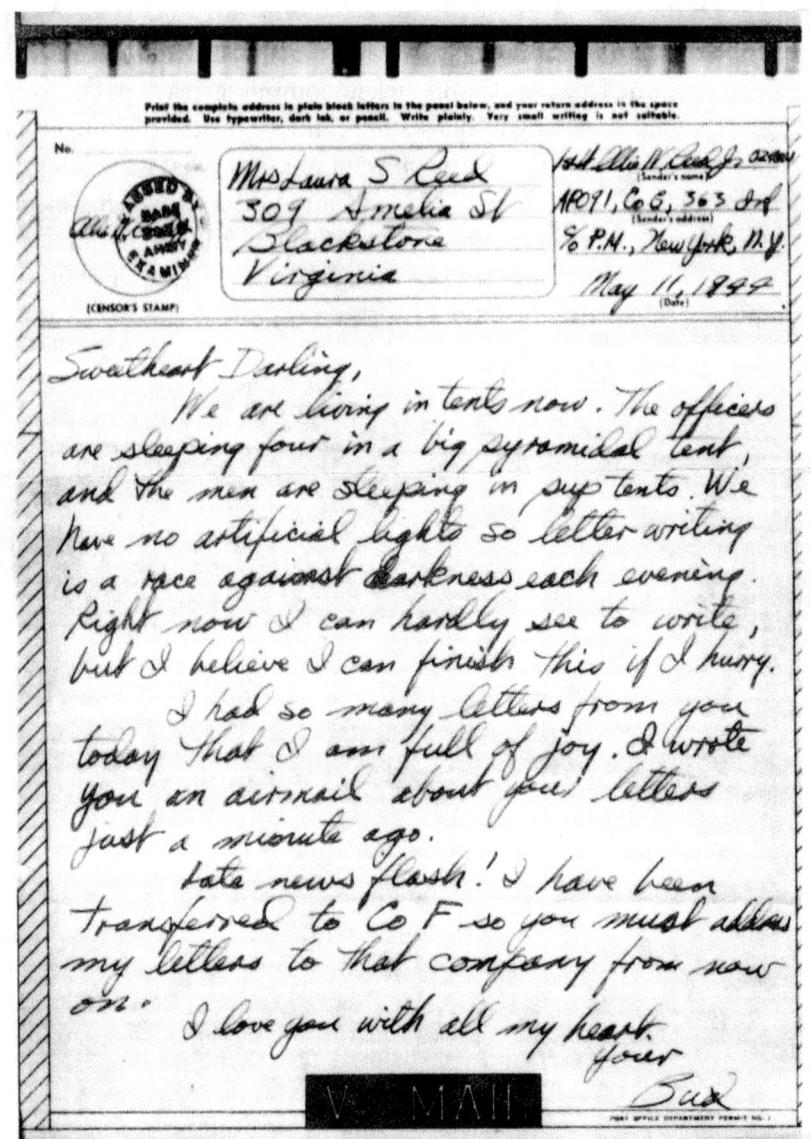

CHAPTER NINETEEN

Safe Landings

91st Division amphibious training landing in Arzew, Algeria. (National Archives)

AS SOON AS HE LANDED on May 11, Bud penned a "Victory-mail" (V-Mail) to Laura letting her know he was on dry land without being permitted to say where.

> We are living in tents now. The officers are sleeping four in a big pyramid tent and the men are sleeping in pup tents. We have no artificial lights so letter writing is a race against darkness each evening. Right now I can hardly see to write but I believe I can finish this if I hurry.

Two days later, he teased Laura with the secret of his whereabouts:

> I had to go to the nearest town today to change all of the company's money from American to foreign. As soon as we are allowed to tell where we are I will send you a bill for a souvenir. I can't send it to you now because the name of the country is on the money.

The Algerian 20-franc note Bud sent Laura.

One week later, Bud sent Laura twenty Algerian francs.

Detail from a letter Bud sent Laura describing the tiny donkeys ridden by local Algerians.

This country is a lot like Texas except that there are a lot of very small trees. The main crop is grapes from which wine is made. The countryside is covered with vineyards. The farms are all owned by French people but the farms are worked by Arabs.

I have visited one big town – Oran. It is very modern in its modern parts and as old as the Bible in its old parts. Most of the people in Oran are French. I will tell you more things as we are allowed to tell you.

Along with being reassigned to Company F under Captain Crowden, Bud also was named defense counsel in the regimental courts martial where trials went late into the night.

By the end of the month, Bud received a letter from Laura complaining that he had not written which was frustrating but invasion preparations put a hold on delivery of letters written after May 21, 1944. In the case of the 91st, mail had been held since they landed, due to their role in Operation Vendetta. He, and the other men in the European and North African theaters, continued writing, not knowing their letters were being delayed for a good reason.

Mildred was worried by the absence of letters from her men overseas. Once D-Day and June 6, 1944 passed, she finally received Ollie's first, very brief postcard, postmarked May 23 – "Everything lovely, Ollie."

This was quickly followed by a May 26 letter:

Dearest: Everything swell

Have traveled by every method except camel back. As you probably know I will not see Bud. All of my traveling has been luxury traveling – everyone has been lovely.

…I am very well but needed exercise when I arrived. We have school four nights each week, so don't have time to get lonesome. Do not have a job – that is a permanent job but expect one soon. Hope you and Ted had a nice trip back and that Ted had a good vacation. It was good to see him – isn't he a fine big man. I am very proud of him. Say 'Hello' to Ollie III and tell him I miss his smile, gurgles and even his occasional yells – at least he was sociable in his yelling. Love to Laura and you two look after each other.

I miss you and love you always.

Ollie

The day after D-Day, Ollie wrote to Mildred of his day in England while war raged only a short distance across the English Channel.

7 June 1944

Dearest: - Well the big day has come & gone. I left here at 0700 6 June for a trip to get acquainted with some base people and it as not until I arrived in the old Roman walled town that I heard the news. It was around 10:00 a.m. Everybody took it quite calmly and very seriously. I visited a beautiful and very old cathedral and added my prayers to the tens of millions that must have gone up on that day.

Am glad to be this much nearer to the real fight but all of us wish we were in it.

Several days later, he added,

In many respects the war seems farther away than it did in the States. I guess the big reason is that we don't have the newspapers, weekly magazines, etc., and that we cannot talk about the war. That last may seem funny but it is absolutely true – having actual knowledge is too dangerous to talk – so we all keep our mouths shut.

Bud was surprised when Laura wrote him with news that his father was in England. Both Bud and his father thought he would be sent to China.

Knowing mail would be delayed, Bud continued writing. In his June 6 letter, he noted the invasion taking place a thousand miles away:

> Today the invasion of France started. I am very glad to hear about it. Perhaps we will beat Germany this year. I pray that the invasion will be successful. I am still in North Africa so you will not have to worry about me, dear.

In a June 14 letter to Laura, Bud wondered his father had participated in the Normandy invasion.

> I received a letter from you two days ago but due to circumstances beyond my control I had to read it in a hurry and pack it away. I will be able to read it again in a few days. You said you thought my father was in England which surprised me because I thought he would go to China. I wonder if he was in the invasion? I guess you don't know yet.

After letters started arriving and they knew their men were safe, Mildred, Laura, Ollie III, and Duchess Nebbie packed into the family Packard and drove to Manhattan, Kansas, where Ted was studying veterinary medicine at Kansas State University.

The Allies had landed in Normandy four days earlier, but, in the pre-dawn hours of June 10, small landing crafts carrying the men of the 91st Division sailed from the Oran harbor. Tossed about in fourteen-foot waves, many of the men became violently seasick over the course of the twenty five-mile trip along the coast. Sick or well, ramps were lowered and the men made their way through the roiling waters at Arzew Beach. Heavy barrages from five-inch guns aboard Navy destroyers whistled overhead. The 363rd hit the beach well off-course, still facing heavy barbed-wire entanglements along the shoreline. Within an hour, the 363rd cut its way through the barbed wire, and knocked out enemy pillboxes with flame-throwers and Bangalore mines. Their mission was a success.

At the close of the amphibious maneuvers, the 91st Division embarked for Naples, Italy, from the port of Oran in the afternoon of June 16, 1944. Soon, they would face with their most difficult "maneuver" yet – actual combat.

Colonel Reed was assigned to XII Corps at Camp Bewdley in the West Midlands of England. Named for the nearby hamlet of Bewdley, it was the headquarters of the XII Corps, one of twenty-six corps in World War II. They were part of Operation Bolero, the massing of troops in the U.K. for the invasion of the European continent. Operation Overlord was the operational name of the invasion of France that would take place ten days after Ollie's arrival. Select officers of XII Corps had been briefed on the details of Overlord on April 26th. Ollie joined the General Staff in planning Operation Neptune, the landing and inland push.

Camp Bewdley was located atop a hill overlooking the town of Stourport-on-Severn, immediately taking Ollie's mind back to 1922 when he was training at Camp Meade, and the serenity along the Severn River in Maryland. Now, in the quiet of the West Midlands, seemingly far away from war, Ollie passed the time imagining a future life with Mildred on the Eastern Shore of Maryland.

On May 30, Ollie wrote to Mildred of a peaceful evening atop a quiet hilltop overlooking the smoke and mist in the valley below, telling her of hearing a cuckoo bird during a walk in the countryside and how he thought she would enjoy it. The next day would be busy, he wrote, as Lieutenant General George S. Patton, newly installed commander of the Third Army, was coming to address the troops.

Arriving with great fanfare, Patton delivered a blistering speech on what is today the front nine of the Wyre Forest Golf Centre. On this beautiful spring morning, long columns of men marched down the hill from the barracks, counting cadence. Soon the green landscape was a sea of brown uniforms. All of Camp Bewdley was in attendance. Colonels and majors packed a stand adjacent to the speaker's platform, all anxiously awaiting Patton's arrival. As a long black car approached, a captain stepped to the microphone. "When the generals arrive, the band will play the General's March, and you will stand at attention."

Dressed in high boots, helmet gleaming in the bright sunlight,

Patton, accompanied by Major General Gilbert R. Cook, commander of XII Corps, and Lieutenant General William H. Simpson, head of the Fourth Army, marched down the hillside directly to the honor guard, with Patton given the honor of inspecting the full-dress line. The first to speak was a chaplain who delivered an invocation. He was followed by brief remarks by General Simpson and then it was General Patton's turn to speak. As Patton approached the microphone, the sea of men snapped to attention.[170]

Unreservedly, Patton called the men to action, leading with the opening line:

> Men, this stuff we hear about America wanting to stay out of the war is a lot of bullshit…Americans love to fight!

His delivery pulled no punches, and indulged no fears.

> You are not all going to die. Only two percent of you right here today would die in a major battle. Death must not be feared. Death, in time, comes to all men…

> There is one great thing that you men will all be able to say after this war is over and you are home once again. You may be thankful that twenty years from now when you are sitting by the fireplace with your grandson on your knee and he asks you what you did in the great World War II, you WON'T have to cough, shift him to the other knee and say, "Well, your Granddaddy shoveled shit in Louisiana." No, Sir, you can look him straight in the eye and say, "Son, your Granddaddy rode with the Great Third Army and a Son-of-a-Goddamned-Bitch named Georgie Patton!" That's all.[171]

In a June 17 letter from England to Mildred, one can easily see where Bud inherited his romantic streak:

> This is a love letter and I wish I could recapture all the love letters I have ever written, squeeze the essence out of them and send it all to you. I'm all yours sweetheart and have been there for thirty years that my love for you has made sweet; with a little bitter still sweet because the love was always there.

> I have been sitting in the long English twilight thinking of you

and of our years together – thinking of what I could say to convey my true love to you – there isn't anything above the word love to me as applied to you. I have such a lovely gallery of pictures in my mind – my gallery is miles long and lined from top to bottom with Mildred – the gallery in Florence is small and valueless compared to mine. The background of my pictures are made lovely with Buddy, Teddy, Laura, Ollie grandson but they are background.

If it wasn't for you what would I do dear – There ain't nothing – nothing – nothing dear, nothing dear but you.

When I come whistling down the lane again be waiting for me, on your feet, with your smiling eyes all mine.

Love,

Ollie

★★★★

As Bud's 91st Division was landing in the Italian port of Bagnoli, near Naples, his father was riding out a monstrous storm roiling over the English Channel. Ollie wrote to his wife:

Mulberry Bridge destroyed by a storm on the Normandy coastline, June 20, 1944 (National Archives)

19 June 1944

I am very cozily fixed up with a berth in the Captain's cabin – Colonel "Jug" Cornog is traveling with me. He is known throughout the Army – football, polo, tanks, etc.[172]

We know the big big outfit we are heading for but that is all. The move has really been slick - just passed along from one officer to another without any orders in the old Army meaning of the word. With all my equipment I surely must look like an animated Christmas tree. By the way the Army sure has been nice to me in the way of transportation. Always comfortable that's me.

The storm lasted two days, delaying Ollie's departure. Along the Normandy coastline, winds whipped waves as high as twenty feet, sweeping over the rapidly constructed Mulberry harbors put into place immediately after D-Day. The temporary harbors provided anchorage for delivery of men and material to fuel the ongoing invasion of the European continent. The massive structures of concrete and steel, weighing thousands of tons, lay shredded along the beachfront.

The next day, Ollie continued:

> This is Tuesday and know but whether it is the 19th or 20th I don't know.[173] The wind and the waves have held us here for 36 hours and no telling when we will move. I get impatient but remember your belief the "Everything – good, etc," and passes away with faith and patience. Am sitting up on the topmost bridge enjoying the keen wind making music in the rigging. The whitecaps are running their endless race and the cloud shadows come and go dappling the water and changing the color of the balloons. Balloons are everywhere – close by peaceful and serene and quite dignified, for off they group and look like a flock of big seagulls balanced in the sky. It is hard to believe that such a peaceful scene is actually the product of war. The soldiers have started singing so we have quite a combination of sounds up here on top. The wind in the rigging and just this instant "Annie Laurie" on a radio, a platoon exercising and other men rattling their mess kits as the line forms for early chow. Regular ship board life reduced to a very small ship.

Upon landing in France, Ollie was immediately driven to his new assignment as commanding officer of the 175th Infantry Regiment of the 29th Division.

Major General Charles H. Gerhardt

The 29th Division was commanded by General Charles H. Gerhardt, the man who initially led Bud's 91st Division and Ollie's classmate from the Command and General Staff School. In the battlefield, Gerhardt had the same hard-nosed reputation he had with the 91st in training. His "March, Shoot and Obey!" motto was always in his back pocket, but he tailored the motto of 29th, "29, Let's Go!" to his own style, distinctly pronouncing each portion: "Two-Nine, Let's Go!" As pictured in *Time* magazine - a shirtless General Gerhardt atop his steed - Gerhardt was ever the cavalryman. He lived and commanded by the hard charge, "*Toujours attaque*" (always attack), was how General Charles L. Bolte described Gerhardt's methods years later. Perhaps this was exactly what was needed when Gerhardt was given command of a division about to fight its way from the landing crafts at Omaha Beach up the sandy bluffs in the face of heavy fire, but many questioned his command style in the hedgerows of the French *bocage* (the Normandy landscape blending woodland and pastures). The men on the ground, as well as corps commanders, had a somber saying about "Uncle Charlie," that he had three divisions – "one on the line, one in the hospital and one in the cemetery."[174] Men feared the approach of his Jeep, nicknamed "Vixen Tor" after a granite outcropping in Devon, England where a legendary witch lived in a cave by that name. Accompanied by his dog, D-Day, riding at his side, Gerhardt was known to come to a screeching halt to upbraid battle-weary men for not properly buckling the chinstrap of their helmet, or officers for not wearing their rank insignias. He was an aggressive commander for whom "Let's go!" had real meaning. He believed in always moving forward and never back, no matter what the cost.

Ollie reported to the headquarters of the 29th Division in the tiny hamlet of Vessie, France, northeast of the crossroads town of Saint-Lô. Gerhardt tersely greeted Ollie upon his arrival: "Welcome aboard Colonel Reed." With formalities over, Ollie was driven directly to his regimental command post, where he took charge on June 23, 1944.

He wrote to Mildred upon his arrival, greatly underplaying what he was seeing, "except for some stuff along the shore and some badly knocked about houses, everything looks fairly nice. Right now I am sitting in a folding wooden theater chair in a quiet orchard

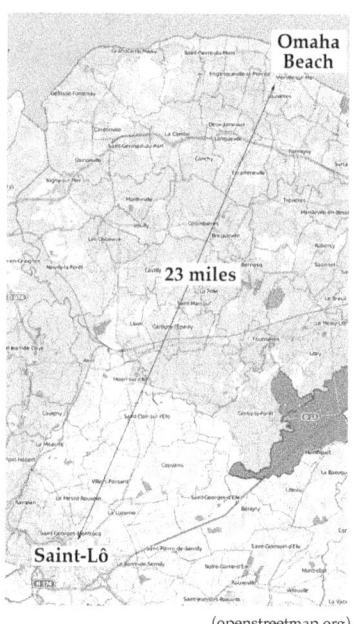

(openstreetmap.org)

hilltop and a nice little breeze is ruffling my few remaining hairs."

In a short time, Ollie was feeling the weight of the command responsibilities he had been assigned, as he wrote home:

> Monday the 26th
>
> Dearest
>
>> This is my third day in command. I have learned what real prayer is – prayer for my men and myself.
>
> Join me – please.

By his side was the book Mildred had given the entire family, *Strength for Service to God and Country*, four-by-five-inch hardcover pocket book. First published in 1942, *Strength for Service* was a compilation of scripture readings for each day of the year, underscored by with a lesson about the passage.[175] The booklet remains in print today. In each of their copies, Mildred had circled her birthday, along with those of Ollie, Bud, Ted, Laura, and Ollie III to add special meaning to that particular day's reading.

Ollie's new position was among the toughest he had ever faced, and could be quite isolating. "I have the feeling that 'family' feeling or friendship means little here," Ollie wrote.

As Ollie would also learn, a replacement officer was no different from a replacement private on the line. In Normandy, he was replacing his old friend from the 29th Regiment, Colonel Paul R. "Pop" Goode who had been captured by the Germans.

Colonel Goode, a 1917 West Point graduate, landed at Omaha Beach in command of the 175th Regiment, was taken prisoner on June 13 while leading a company on an assignment from General Gerhardt to capture a bridge and round up stragglers from an earlier unsuccessful attempt.[176] Goode was certain the general was ordering a suicide mission and refused the stay behind after ordering his men to march into certain peril and refused to not accompany them. Goode told Captain Kernan Slingluff, commanding officer of Company K, "Captain, I wouldn't order anybody to go into a thing like this unless I went myself, so I am just along for the buggy ride. Just consider that I am not here, and I will go along."

Slingluff led his men deeper behind German lines where they engaged the enemy at the bridge over the Vire River. They were badly out-gunned and out of range of artillery support. The platoon to his left started with forty-two men and one officer, but was already reduced to five and a wounded officer. Slingluff suffered shrapnel wounds from a German mortar to his knee and hip, "but it wasn't bad," he told his father in 1945. Finally, a rifle bullet pierced his right hip, making it impossible to move. As Slingluff faded in and out of consciousness, his company risked encirclement. He ordered his second in command to gather the men and fight their way through the right side, leaving behind anyone who could not walk on their own – and that included him. His first sergeant threw him a half bottle of cognac, saying: "Captain, you always said you wanted to die drunk."

Colonel Goode had not been hit, but he refused to leave, saying that he had come to see the operation through. Exhausting all of their machine gun and mortar ammunition, and with little rifle and carbine ammunition remaining, Goode made the decision to stop the fight.[177]

"Captain, they have killed enough of your men now. I am going to surrender you."

Slingluff ordered his men to cease fire, and Colonel Goode stepped out and surrendered the remains of Company K.[178]

Thirty-five men managed to escape capture and eighteen were taken prisoner, meaning that only fifty-three survived out of the 225 who started the mission.[179]

Goode was taken prisoner and held in Stalag 7a in Moosberg, Germany until the camp's liberation in April 1945. Slingluff was held in Poland and walked away from the camp when the Germans abandoned it in the face of oncoming Russian forces.

Goode was succeeded by his executive officer, Lieutenant Colonel Alexander George, a 1920 graduate of West Point, who, in turn, was severely wounded on June 17 in the advance on Hill 108. George suffered severe wounds from a German grenade blast as he was leading a First Battalion patrol against an enemy machine gun position. George had already earned a reputation for recklessness. Following D-Day, George found a discarded bicycle and, as GIs watched in stunned amazement, he rode along the front lines yelling, "Give 'em hell, boys!"

"He sure as hell didn't look the part of an executive officer, let alone a West Pointer," recalled one infantryman. George also took charge of a rifle squad in a successful attack against a troublesome German "88," The feared German artillery weapon being used against tanks and infantry.

The report of his wounding reached General Gerhardt.

"He was leading a patrol against a machine gun nest and someone lobbed a grenade, and it got him full in the face," Major Leslie Harness, the 175th's S-3 officer, reported to Gerhardt over the telephone.

"He shouldn't have been up there," Gerhardt snapped.

"He was hit above the eyes and his nose is half gone and he has holes in his back," Harness replied. "He'll be a casualty for some time."

George survived but was replaced by Lieutenant Colonel William Purnell, a Maryland National Guardsman and longtime member of the regiment, who would eventually rise to the rank of major general. Purnell's command was only temporary as Gerhardt demanded a professional soldier from the Regular Army to take over the 175th, and that would be Colonel Ollie Reed.[180]

The 175th Regiment was in the fight for Hill 108, named "Purple Heart Hill," for the number of casualties suffered in the fight for the strategic high ground. Colonel Reed was given command of the 175th Regiment in the midst of the battle to control the hill and move forward. Purnell stayed on as Reed's executive assistant.

When Ollie took command, the 175th had been in non-stop combat since landing on Omaha Beach more than two weeks earlier. The regiment successfully opened the advance on the first major goal, the crossroads town of Saint-Lô, on June 16–17, pushing the 29th Division well ahead of the 30th Division. Elements of the First Battalion advanced within two miles of Saint-Lô, making it appear the regiment could take the German-held town on its own. General Gerhardt was delighted, reporting to General Charles H. "Cowboy Pete" Corlett, commander of XIX Corps, "I feel we'll be getting to Saint-Lô before long."

Instead, the advance was bogged down in the thick hedgerows of the Normandy countryside, which provided cover for German infantry, tanks, and mortars. The 29th was forced into a more defensive posture, against General Gerhardt's instincts. Some feared the situation could quickly evolve into a stalemate along the lines of trench warfare of World War I. In Normandy, the fighting would be from hedgerow to hedgerow. Ollie wrote in the gallows humor of life on the front lines:

Hedgerow fighting, Normandy. (National Archives)

> Before I forget I want to tell you of the chuckle I get three times a day at least. We (American soldiers) move around with a great deal of respect for the German shell fire – our own guns shoot over us and frequently we'll dodge first and then realize that that shell is going out – not coming in. A sudden dive is normal and quite uncalled for. As I went to breakfast the other a.m. I took such an uncalled for dive and twenty yards later turned a wall corner to see an old 80 year old grandmother calmly plodding across the barnyard with a chamber mug in one hand and her staff in the other hand. There are two French families within 400 yards of my CP calmly going about their normal business in life. There is also a little grey jackass in the field (the fields are about 200x200 yards) next to mine who has gone through the shelling with never a duck or a scratch and gives his triumphant bray morning and evening. If he loses his rabbit foot I will miss him a lot…Am mainly concerned now with rejuvenating, refitting and raising the morale of this outfit.
>
> Had a pleasant surprise today. A big grin and 'aren't you Colonel Reed?' One of the 309th boys – have been disappointed not seeing more of them.

Frontline replacements were often shocked by the sight of combat veterans. A young lieutenant from the 35th Division recalled his rifle company replacing one from the 29th.

Members of the 175th rest in a Tessy sur Vire taproom.
(Maryland Military Museum / National Archives)

We had little idea what the 29th had been through. We found all the men wearing their field jackets reversed. It reflected a lot of light. They had [them] turned inside out because on the inside was a kind of dull lining and they were trying to get the effect of camouflage. Officers didn't carry anything that would mark them as officers. Their bars were all concealed. They discarded a lot of their equipment. They were a pretty badly beat up outfit. We hadn't expected anything like that, and our first reaction was that this is not a very good outfit.[181]

By late June, when Ollie took command of the 175th, an "old" soldier was one who had survived at least three days on the line. Battle fatigue was setting in. Some men aimlessly staggered around the front like wide-eyed drunks, while others adopted the hardened one thousand-yard stare. Medics dispensed blue tranquilizers that jokingly became known as "Blue 88s," as powerful as the German 88-millimeter gun.

Casualties steeply mounted. The 5,211 members of the 29th Division rifle companies, thirty-seven percent of the division's total manpower, suffered over ninety percent of the division's casualties. Most 29th Division rifle companies that landed on D-Day had a near-complete turnover in personnel by mid-July.

By the time a replacement arrived, his morale was already shaky. He had been dislocated from his training unit and suddenly dropped into a holding area on the edge of a war zone filled with the sights, sounds, and smells of combat, seeing the dead and wounded coming back from the front. The replacement was something of an orphan. He knew no one and no one wanted to be his friend. Combat veterans had already lost enough friends and they did not need any new ones. "I have seen men killed or captured when even their squad leaders didn't know their names," one infantryman said.

Joining a new unit in the midst of battle was the worst. "I saw replacements headed for an infantry regiment brought in under the cover of darkness," Lieutenant Colonel John P. Cooper, commanding officer of the 110th Field Artillery, recalled. "By midnight they were in a foxhole and at 5:00 a.m. they had to attack. By 5:30 a.m., some of them were coming back on stretchers - dead."[182]

With what he had seen over the past month, Ollie did not want Ted, a private in the reserves at Kansas State, to step into the action, and told him so in a July 26 letter:

Ted Reed, 2nd from right, sworn into the student reserves.
(Kansas State University Library and Archives)

> Yes, I certainly approve of your actions in continuing your Vet studies instead of going into the Infantry. You will be much more valuable to your country as a Vet than you would be as an Infantryman. There are plenty of men who are not good for anything but being Infantrymen. You should do the hard and specialized work that you are fitted for, and that few other people can do.

To Mildred, Ollie was emphatic about Ted sticking with school:

> We came out this a.m. and the Good Lord gave us a covering of fog that lasted until the unusually late hour of 1130 a.m. I really feel that the Lord gave us of his mercy. We wrote what will be referred to as a "heroic page." The boys are heroic - but we are not professional killers and most of them never will be. Keep Ted out of it so long as can be w/o hurting his own self respect.

Ted was enrolled in the veterinary medicine program at Kansas State University, and a member of the campus reserve unit. As a group, the veterinary students trained together under the supervision of a second lieutenant. On one occasion when they were marching poorly, the lieutenant in charge brought them to a halt, chewing them out.

"You are a company of veterans," the lieutenant screamed. "You're not marching like veterans. You're not behaving like veterans." He was interrupted by one of his charges, "Sir, we're not veterans, we're veterinarians."[183]

Between the end of June and the first week of July, the 29th Division was given a chance to rest behind the lines. They had been in constant combat since June 6 and needed a break. The men enjoyed hot showers, warm meals, and entertainment, such as movies, in the relative calm a mile or two behind the fighting front.

On July 2, Generals Eisenhower and Bradley visited the 29th Division headquarters, a mile southwest of St. Clair-sur-l'Elle. Ever the showman, Gerhardt welcomed his commanders with the division band blaring military marches, no doubt heard by the Germans. Wearing clean uniforms belying the rain, mud, grit, and grime of daily combat, the majority of Gerhardt's regimental and battalion commanders were pulled from the front lines to stand at attention outside the divisional war room. Gerhardt introduced each of his commanding officers to his distinguished guests as he accompanied them along the greeting line. Bradley's aide, Major Chester Hansen, noted in his diary that the 29th was the only division in Normandy that had greeted his boss with a formal ceremony. Surprised that Gerhardt had brought so many field commanders from their front line posts, Bradley joked to Eisenhower, "What would have happened if a war had started on the front?"[184]

As Ollie reported in a July 5, 1944 letter:

> Can't remember if I told you that Generals Eisenhower and Bradley visited us recently? Bradley stopped and chatted bit – he is certainly an easy, nice person.

General Omar Bradley fires an artillery round celebrating the 4th of July in Normandy. (National Archives)

> As if we didn't have enough noise around here and not enough ammunition, 1st Army HQ ordered a grand 4th of July celebration. At the stroke of noon, every weapon in Normandy opened up – from sidearms to cannons. It was quite the spectacle.
>
> By the way a sudden stoppage of mail from Buddy can be expected if they ever start to move – so don't be alarmed. "Security" is the answer and of course security is well worthwhile.

CHAPTER TWENTY

Italy

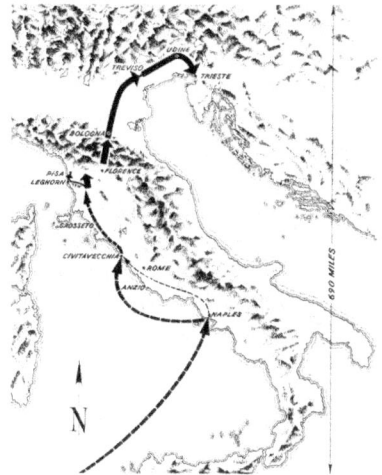

Route of the 91st Division in World War II.

WITH THEIR AMPHIBIOUS TRAINING COMPLETED, Bud's division boarded transports for the trans-Mediterranean crossing from Oran to the Italian port of Bagnoli and two weeks in Naples as they awaited supplies for the northward march to join the fight.

The Italian mainland had been breached by combined American and British forces on September 3, 1943 in Operation Avalanche. Sicily had already fallen to Allied troops in August when German forces fled to the mainland. The Italian government dissolved days later. Dictator Benito Mussolini had been voted out of power on July 25, 1943, by the Italian parliament. He was immediately arrested and imprisoned, only to be rescued by German special forces, led by Otto Skorzeny.

Intense battles inflicting heavy losses on American, British, Canadian, and German units preceded the march of American troops into Rome on June 4, 1944. The city had been abandoned by the Germans and declared an "open city," allowing Allied troops to enter without much resistance. General Mark Clark, commander of the Fifth Army, construed orders to his own liking, entering the city, rather than a flanking action as ordered by the British commander of Allied forces in Italy, General Sir Harold R. L. G. Alexander. A popular reasoning attributed to Clark's decision is that he wanted to have his picture taken in Rome as a conquering hero.[185]

On June 19, 1944, Bud wrote to Laura about his new location:

Last photo of Bud in Italy.

I just got the news on what I could tell you about our latest move. We are in Italy but I can't say where nor mention any towns at all like I could in Africa. I have passed through some towns tho and I have seen the effects of bombing in them. The first marks of war I have seen. I am living in barracks now which is much drier and cleaner than living in tents. Those are the only facts I am allowed to tell you except I can tell you about the people and what they do.

The Italians seem much better off than the North Africans but they are far below American standards. The people are just like the Italians in New York only their clothes are generally in poor shape. There does not seem to be a great shortage of everything as in Africa. There are plenty of fruit stands open and some other stores too. The people are poor. Any American soldier is a rich man here. The morale of the people seems low judging from the prostitutes which are everywhere and very conspicuous. They are very bold – in fact, as I was walking down the street two different women grabbed my arm and tried to stop me while they told me their price which was fifty cents American money!

The kids are about like the Arab kinds in the way they follow soldiers around hollering "Hey Joe, gimme cigarette, gimme candy, Hey Joe!" Many of the kids know American songs such as "Pistol Packin' Mama." They seem much more intelligent than the Arab kids but they are all barefoot and ragged and dirty. They know all the American slang and vulgarity too.

The country is much more hilly and much, much more greener than Africa. The hills are small but very up and down. Walking is tough because of this. They have very funny trees here that look just like umbrellas.

Also on June 19, Laura wrote to Bud about her daily routine while living with Mildred in Manhattan, Kansas. She placed a sticker of flowers in a basket with the note, "Smell your June flowers, dear!"

My Dearest,

I had a nice quiet day for a change. I was not hurried so could rest easily.

The baby woke me up about 7:00 but I gave him Time magazine to tear up which he did without delay. I then got up about 8:00 and cleaned the baby up – and gave him some food and he took his morning nap.

After he got up at 10:00 I carried him across the street to where mom was having her hair done. We stayed in the beauty parlor about an hour. Then came down a half block to the shopping district. We looked for an hour for a carriage and had given up and started looking for sun suits for the baby - when we went into Coles department store and they had four baby buggies. I picked a black one with the most metal – and rubber tires. I was then so glad to set Ollie down that I bought (Mike will pay for it) a sheet and mattress and rubber sheet. Then I put him in it. He loved it and just hooted and laughed.

Laura, Ollie III and the new stroller.

We then rolled ourselves around the block and got a bottle of milk for him. Then went out right away again. Ollie drank some milk then went to sleep. We shopped for a birthday card for Pop which I got and signed our name to. We found a baby swing which we can hang on a doorway for him in a month or so.

Mom and I got back at 5:00 I woke the baby and gave him his much loved bath then put him on the bed – which I have told you he rolls around on. Well tonight he rolled on his tummy and crawled over to his little toy. Isn't he wonderful?

> Then Ted arrived and we went to supper – Ollie in his carriage asleep. This evening I washed out the clothes, i.e. diapers, etc. And took D.N. [Duchess Nebbie] for her walk – I am now undressed about asleep.
>
> I miss you so very much dearest one.
>
> Forever I remain your Sweetheart

Laura came up with a notion that she would to take flying lessons and join the Women's Air Corps so she could be stationed close to her husband. Bud wrote with his support, but cautiously.

> Well, sweetheart, I am surprised that you really and truly want to enlist. I have thought about this hard since I received your letter and I feel this way about it. You have never been wrong about anything important since I've known you. I have great confidence in you and I know that whatever you do will be right. I will always back you up as strongly as I can. Therefore I will send you my written permission to join any branch of the armed forces in this letter. In many, many ways you will be doing a fine thing and you will be serving your country in time of need. There is one very important thing that I am afraid of and that is that the Army (or Navy or Marines) will change you – make you harder and destroy the thing I love most in you – your inner loveliness.

> I am looking at the snapshot now of you and our baby, as you seem the sweetest and loveliest woman in the world. You are the wonderful person who satisfies me with her love – who brings me happiness and peace. If you change it will break my heart and our marriage too.

On June 24, Bud mailed a "To whom it may concern" letter, witnessed by Second Lieutenant Joseph W. Bone, on Laura's behalf with his permission for her to join the armed services. He added a warning about the harshness of Army life:

> You remember the words you used to ask me the meaning of? In

Love and Sacrifice

the army you will find women who use those words in every sentence they speak. You will find hard-drinking women and loose moraled women. Every soldier and officer you see will be trying to "make" you because the WACs and WAVEs have a terrible reputation among the men.

Yes, you could talk with me all right. I've seen plenty of officers dating WACs and heard a lot of stories about it too. Only the chances are 100 to 1 against us meeting over here because the army covers the whole world now and you or I may be sent anywhere.

In late June, Bud was shocked to receive a letter from Laura addressed from Lexington, Missouri. Mildred had taken Laura to Lexington, showing her around the town and introducing her to people Mildred knew. Although Nancy Campbell was away at college, her name certainly came up during the visit. For Laura, a young, insecure wife whose husband was far away on the frontline, such an encounter with Bud's teenage years and the specter of Nancy as Bud's true love, was a dagger into Laura's heart. Fearing that Nancy was the woman Mildred wished was her daughter-in-law must have been painful beyond imagination, causing her to send a note to her husband overseas, writing to him during her trip with her mother-in-law to Lexington. Bud replied on June 26th:

Your letter from Lexington interests me a lot and I want to talk to you about it. I know a lot of the town bored you and you probably got pretty tired of the houses and places that I have lived in and been at. I don't blame you but I know you were nice about it all anyway. Mom was really a big leader in that town and so was Pop. He was head of the Masons and a big man over at Wentworth too. The people there really liked them both and they will remember them a long time. But I, myself, didn't have much to do with the place and I can hardly remember anyone there. I know the Ardingers and the Sellers and that's about all. So don't you think you have to like the town or get to know the people or anything for I don't care if I ever go back or not.

And now about this Nancy business! She has a lot of friends there too so don't believe that baloney put out about her. You are nicer and sweeter than she ever was. Now don't you forget that for a minute! Nancy was very sweet especially when I first knew her but she got harder and harder as time went by and she didn't even bother writing me after a while. So she didn't end up as sweet or as nice and she didn't have her heart broken either. I thank my lucky stars that I married you, lover, because you are worth a million Nancys. I mean it, dearest Laura – you are the only one I've ever loved so much and you are the only one for me. I like the way you end your letter by saying that you were the one that got me and that's what counts. You are 100 percent right and I'm really glad you know that and don't let these Lexington people worry you. I love you and you alone.

Honey, you always looked to be older than you are. Maybe because you are "well-developed." I love you that way and I always will. The Sodingers are just used to those skinny Missouri gals so don't worry about their age-guessing!

Well, dearest, that's enough for Lexington and everything connected with it. You are my wife and you are my lover. My "home" is where you are. Tell me that you believe me and that you love me and that you'll forget about this town. I knew you'd have to go there some day and I'm glad it's over with.

Don't forget – my home is in your heart.

The 361st Regiment of the 91st Division entered combat in advance of the other units, relieving parts of the 34th Division that had participated in the January 1944 invasion of Italy. After landing at Anzio on June 1, the 361st fought their way up the western coast of Italy, encountering heavy losses, particularly at Casole d'Isola, where the fighting lasted four days, July 1 through 4, 1944.

For Bud's 363rd, the sunny respite of Naples was coming to a close. In the days before the 91st Division moved from Bagnoli, battalion commanders, with their staffs, accompanied by platoon leaders, including Bud, and their sergeants drove inland toward the port city of Civitavecchia, through the streets of Rome, past Monte Cassino, and Anzio.

The regimental headquarters of the 91st Division was attached

Love and Sacrifice

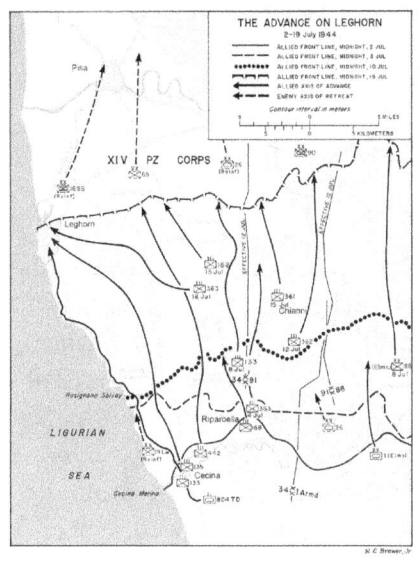

From *US Army in World War II*, amended.

to the 34th Division under a plan where the battle-hardened veterans of the 34th would lead the 363rd Regiment into their first combat, with Colonel W. Fulton Magill in command. The 34th had been in the fight since 1942 in North Africa. They landed in Italy at Anzio Beach, behind the initial invasion, and fought their way up the coast.

The bulk of the 363rd arrived at Civitavecchia via landing crafts on July 2, having sailed overnight along the western coast of Italy from Bagnoli. The men were immediately shepherded from the LCIs (Landing Craft Infantry) to waiting troop trucks. They were eager and boisterous, especially when they saw the charred remains of German equipment along the northward ride toward the town of Cecina. As they neared the Cecina River, the men caught their first sight of German artillery airbursts. An immediate hush fell over the group.[186]

Civitavecchia harbor, 1944 by Ludwig Mactarian (US Army)

The approached to the bridge over the Cecina River leading to the town of Cecina, was backed up for miles as elements of the 34th Division crowded the road into the city that had just been captured. Plans were immediately changed, moving the 363rd to an assembly area on the north bank of the Cecina River, west of Casaglia. Word of the change in orders and direction had to be spread through every method possible to direct men and supplies of both the 34th Division and 363rd Regiment

to the new position. On the night of July 3, the 363rd Regiment fell under their first artillery fire, and orders were issued at three o'clock in the morning for the combat team to move northward at daybreak.

Company F, part of the Second Battalion, was assigned the objective of the high ground overlooking the town of Chianni.

Accomplishing that mission, they would jump off toward the port city of Livorno, popularly known as "Leghorn." American forces had already driven the enemy back one hundred-fifty miles, to the outskirts of the port city, but the Germans intended to defend the approaches to Livorno, the third largest port in Italy. To avoid the heavy losses of a direct attack on Livorno, the plan was to employ a flanking movement. This tactic avoided the grinding door-to-door and alleyway combat that had cost many lives and flattened towns in the advance from the south. Instead, the port city would be bypassed and isolated, leaving it vulnerable to capture without a bloody fight.

This required American forces, including the 363rd Regiment, to swing inland and proceed north, rather than hugging the coastline. The terrain was mountainous and ill suited to rapid, mobile military operations, but casualties would be reduced. With success, two primary objectives would be accomplished: the capture of the port of Leghorn and control of the Arno River. Afterward, combined elements would continue pushing north for an assault on the German defensive stronghold known as the Gothic Line.

The area between the Tyrrhenian Sea and the Tiber River is a contour of numerous isolated groups of hills and mountains, many volcanic and rich in minerals, particularly mineral-rich limestone rock. The inland path the 363rd would follow took them along a north-south range of hills with a rising ridgeline, a veritable stairway of hilltops, as they moved northward from the Cecina River. The 363rd had a series of hilltop objectives, most without names, only elevations:

363rd Regiment marches up the hills above Cecina, Italy. (National Archives)

Hill 504, Hill 457, Hill 506, and Hill 507, plus named promontories such as Mount Vitalba, Montalone, Pogo Prunicci, and Rostona Ridge. Overtaking Hills 553, 505, and 477 west of Chianni would give the 363rd strategic domination over the town of Chianni, an initial objective.

On July 4, 1944, when the 29th Division in Normandy was firing every one of their weapons into the air in celebration, the 363rd Regiment entered their first combat near the town of Riparbella.[187]

That morning, Bud penned a letter to Laura:

July 4th

My Laura,

I thought that I would have plenty of time to write this letter. I wanted to try to tell you how very much you've meant to me. You are my life. That is the only way I can tell you how much you mean to me. You are my life.

I write this early in the morning sitting behind a haystack in a field. In a half hour I am going to work. I do not know when I will be able to write again. I will think of you every day and pray for you each night. Will you do the same for me?

I have loved you from the moment we met and I will love you to the end of time. You are always in my heart.

Your husband

Bud

The 91st faced small German combat patrols of a badly disorganized units of the 19th Luftwaffe Field Division (Luftwaffen-Feld-Divisionen, LFD), under the command of the XIV Panzer Corps. The LFD infantry units had been formed by Reichsmarshall Hermann Göring, chief of the German air force, the Luftwaffe, in an effort to broaden his reach into the Führer's military campaign.

The First and Third Battalions of the 363rd led the advance, while the Second Battalion, including Bud's Company F, was held in reserve.

Shortly after noon, the Third Battalion began their attack on what was designated as Hill 4 in the initial engagements. Four and a half hours later, Hill 4 was taken without casualties, while the Germans suffered twenty killed and eleven captured. As evening fell, a company was sent to the right flank of the Second Battalion to prepare for the next day's action.

The 19th LFD was becoming more disorganized with every engagement, but they were supported by the Sixteenth SS-Panzer-Aufklärungsabteilung (16. SS-Panzer-Grenadier-Division "Reichsführer-SS"), commanded by Major Walter Reder. This rather rabid division had grown from Reichsführer Heinrich Himmler's personal escort battalion, and went on to commit numerous atrocities in pursuit of Italian partisans over the remainder of the war.

At six a.m. in the morning of July 5, the Third Battalion suffered their first casualties as they led that morning's advance. Three men were killed and four wounded in an artillery barrage. They continued to move under small arms fire, slowly advancing five hundred yards.

The Second Battalion of the 363rd was visited by the regimental commander, Colonel Magill, and shortly after 7:00 a.m., the Second Battalion moved out with Company G at the lead.

Units of the Third Battalion were pinned down by snipers and small arms fire. In the fog of battle, the Second Battalion's radio operator was killed by friendly fire, accidentally shot by a sentry.

Before 8:30 a.m., the Third Battalion put out a call for an ambulance from the 316th Medical Battalion as casualties were mounting. German snipers started firing on the medics, who were clearly wearing the red cross insignia of the medical corps. The Second Battalion was ordered to send platoons from Hill 553 to the Third Battalion's right flank. They immediately met small arms and machine gun fire slowing their advance.[188]

At 11:45 a.m., Company F, under the command of Captain Eugene E. Crowden, moved five hundred yards to the north and east on the right flank of the Third Battalion to secure Hill 519, the source of heavy fire.

Their next move was north along the ridge to anchor Hill 506, a small hill southwest of Chianni, and continue protecting the exposed right flank of the regiment. As Companies E and G engaged in a firefight, Company F was also meeting strong resistance. The enemy was entrenched four hundred yards north of Hill 506, under the cover of a group of farm buildings on the forward slope. A little after 1:00 p.m., as Company F secured Hill 506, the Germans unleashed a furious artillery fusillade. Lieutenant Reed ordered his panicked men to cover, but some started running. Shrapnel from a 177mm round hit Lieutenant Reed in the neck, killing him almost instantly.[189] [190]

By 6:00 p.m., Company F had taken the farm buildings that had been the source of much of the small arms fire. Two platoons dug in on the north side of the house that had been badly damaged by artillery fire, and a machine gun was put in place. The command post for Company F moved back to Hill 506, where Captain Crowden called in his report that fifteen Germans had been killed, another three wounded, and no prisoners taken. Without specific numbers, Crowden reported that the company had lost one half of one platoon and two-thirds of another. There was no mention of Lieutenant Reed in the initial report.[191]

Bud and four members of his platoon were left lying where they fell, covered by whatever their comrades could muster, on the rugged ground of Hill 506, while the 363rd continued their rapid pursuit of the retreating Germans. All five men were initially reported missing in action.

As Captain Crowden wrote to Laura from the Italian front in August 1944:

Captain Eugene E. Crowden

> It was on July 5 that we were ordered to take a small hill in the mountains southwest of Chianni, Italy.[192] We took the hill no sooner gained our objective than the enemy laid down one of the heaviest concentrations of artillery we have yet witnessed. Several of Bud's men started running instead of hitting the dirt and trying to get cover. While Bud was attempting to get them down, the fatal shell fell, killing Bud and several of his men instantly. The skirmish lasted only two hours, but none of us will ever forget its intensity.

The next day, Captain Crowden led Company F up the east side of Hill 553 where they engaged in an all-day firefight. The 91st Infantry Division went on to storm the Gothic Line and punch their way through some of the heaviest and toughest fighting in Italy and in World War II.

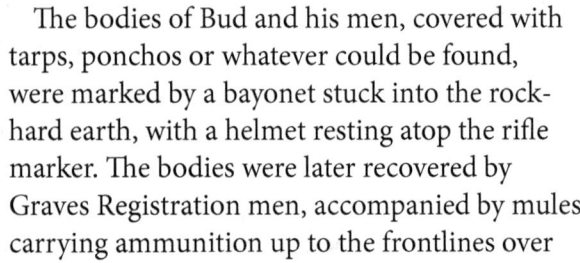

The bodies of Bud and his men, covered with tarps, ponchos or whatever could be found, were marked by a bayonet stuck into the rock-hard earth, with a helmet resting atop the rifle marker. The bodies were later recovered by Graves Registration men, accompanied by mules carrying ammunition up to the frontlines over the rocky terrain, and down bearing the bodies. Mattress covers were unrolled as the Graves men wrapped each body, tying each cover at the end, with the soldier's dog tag attached. The pack animals carried the men, slung over their backs, to headquarters and then their temporary resting place in Vada, Italy.[193] There, Bud laid at rest between two of his comrades from the 363rd Regiment: Private First Class Russell J. Johnsen, Company F, from Minneapolis, Minnesota; and Staff Sergeant Arlyn R. Turner, Company C, from Chicago, Illinois. Also buried with them were Company F casualties from 5 July 1944: Private First Class William Noah Baker from Dickson, Tennessee; Private First Class Gathier Howard Cain from Fayetteville, North Carolina; and Private First Class William L. Morris from Barterville, Kentucky.[194]

Among the items sent home to Laura was a blood-stained pocket case of photographs, a chess set, *Strength for Service*, Bud's wedding ring, and his West Point class ring, from which the gemstone at its center had been pried from the setting. Today, his son wears that ring, with the stone still missing.

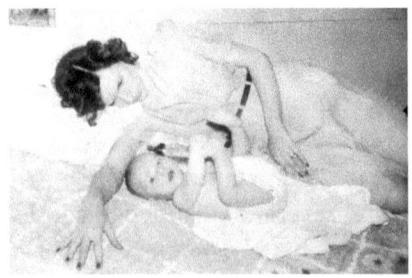

CHAPTER TWENTY-ONE

Normandy

R&R behind the lines in a Normandy apple orchard. (National Archives)

HAD A REAL DOWNPOUR YESTERDAY. I remember Hall, Cope and Gertie in high school days singing, "When it's Apple Blossom Time in Normandy." Such a romantic picture and what a laugh compared to the actual scene! That song has been ruined for me. When I hear it, it will bring anything but pleasant memories.

> When it's apple blossom time in Normandy
> I want to be in Normandy
> By that old wishing well with you Marie
> When it's apple blossom time in Normandy
> I'm coming back to you
> And the spring will bring a wedding ring
> Little sweetheart to you!

The artillery is playing hell with our sleep. At first, it was the Germans but now our own artillery has become top dog, keeping us awake. When they cut loose, it's a combination of lion's roar and the clearing of a tremendous throat. Very reassuring.

In the same July 9 letter, Ollie wrote about getting his first bath since arriving in France, closing with praises for the men he was leading.

Everybody literally marvels at the fortitude of these kids – especially when wounded. Courteous – don't want to cause any trouble – grateful for everything and never a whimper.

The next day, July 10, Colonel Reed was awarded the Bronze Star by General Gerhardt:

For meritorious achievement in combat as regimental commander in Normandy, France. From 23 June 1944 to 10 July 1944, Colonel Reed directed his Regiment in the difficult actions that characterized the early stages of the Normandy Beachhead. In the face of virtually insurmountable obstacles, he inspired his command in the destruction of many enemy hedgerow positions and the gaining of important objectives. His courage, initiative, and superior leadership reflect great credit upon himself and the Military Service.

That day, Ollie wrote to his wife:

I have had a great day – six letters from you and a cablegram. Brings you up to June 20 to me. Those letters will be read and reread. Thanks, darling. By separate cover I am sending along a bronze star and the order awarding it. Undeserved, but am pleased to have it. Will write more later, but have a lot of sleep and dreams of you to catch up on.

The final assault on Saint-Lô began on July 11. The 175th Infantry Regiment would push south and west of the main assault on Martinville Ridge. General Gerhardt ordered Lieutenant Colonel Purnell to move the Second Battalion forward to join the 116th Infantry Regiment on Martinville Ridge. "The other battalions will move up under cover of darkness so that you will be set to go through the attack in the morning," Gerhardt assured him. "Have Colonel Reed call me when he gets a plan in his head."[195]

Martinville Ridge provided a high point over the Saint-Lô–Bayeux highway and within sight of their objective: Saint-Lô. As the attack unfolded, the Second and Third Battalions of the 175th moved to secure the Saint-Lô-Bayeux Highway and cut off German supply lines, and open a direct path into the city of Saint-Lô. Colonel Reed devised a plan of attack utilizing two of his three battalions in attack columns. This suited Gerhardt as a column attack against a line was an approach straight out of their days at CGSS when they studied the Napoleonic Wars. Napoleon was another general who loved continuous forward momentum, like Gerhardt. Led by mechanized units of the 747th Tank Battalion, the attack would resemble a cavalry column breaking through the line, dividing and scattering

Martinville Ridge (Center for Military History)

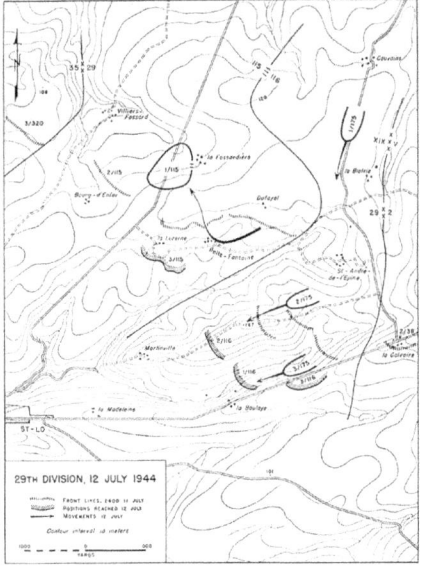

the enemy. The realities of the hedgerows were another matter, no doubt, leaving General Gerhardt longing for his horses.

In the calm of an early morning mist, artillery and mortars rained on the 175th from a parallel ridge to the south. German observers had a clear view of the battalions marching along the hedgerow-lined road toward them. All efforts at progress off the main road were thwarted by a continuous line of hedgerows bordering both sides.

Hedgerows, typically, are six to ten feet wide, and five to six feet high, covered in thick brush and trees. Underneath the surface is a tangle of roots and debris accumulated over hundreds of years by farmers moving refuse from the lands to the sides of local roads, on top of which the heavy cover grew over time.

In the course of four hours, the Third Battalion had moved five hundred yards while the Second Battalion suffered heavy losses, realizing little gain. Ollie tried everything in the book to break the

Tank with bulldozer attachment breaking through a hedgerow. (National Archives)

German grip, but even his use of engineers and a bulldozer to cut an alternative route, out of German view, was thwarted by German artillery. The work of carving a four hundred-yard path through the thick terrain had to be done by hand.

American artillery fired white phosphorus rounds along the ridge without successfully stemming German fire. Colonel Reed asked that his remaining battalion, the first, be put into service for protection along the flank of the main attack, but his request was denied by General Gerhardt, who refused to risk his last reserve battalion. The Second Battalion swung to the southeast of the Third Battalion, broadening the attack, but this move netted only one hundred yards by the end of the day.

On July 12, Gerhardt was beside himself trying to get the 116th and 175th to launch an aggressive assault down the Martinville Ridge toward Saint-Lô. The crossroads town was only a tantalizing two miles ahead, but no one harbored the notion that it would be an easy stroll down the lane to get in. Gerhardt was insistent that an assault down the southern face of Martinville Ridge by the 116th Regiment with the 175th joining them would open the way into the town.

"Let's keep pushing now," Gerhardt told Colonel Charles D.W. Canham, commander of the 116th Regiment that morning. "Keep Dallas's outfit [first battalion] moving now. Get on down and get hold of Reed and get him set to move. We'll never get anything accomplished today unless he moves by noon. The day is short, you know."[196]

An hour later, Gerhardt ordered Reed to advance his Second and Third Battalions through the 116th positions.

"What it takes is driving. You are the guy that has to do it," he told Reed. "Where you are to do it from is your business but we've got to get on those final objectives. Time is wasting but I think your gang is in good shape, so get them rolling. When you report, get facts. There are too many uncertain reports coming in."

Ollie had but one reply: "Yes, sir."

Gerhardt shot back: "Good. Now, let's go!"

As the day unfolded, the 116th slogged through the hedgerows, gaining only six hundred yards. At that pace, the 116th would take another week to capture Saint-Lô but not before running out of men. Martinville Ridge was a shooting gallery. German artillery fire was so intense from the southern ridge that Reed's 175th fought with everything they had just to move into secure positions. His Second and Third Battalions suffered almost sixty casualties in their effort just to join the fight.

Gerhardt was incensed. When he called the headquarters of the 175th at 1:20 p.m., Reed gave him the locations of the Second and Third Battalions. The general lost his temper.

"What are they doing so far back?" the general demanded of Colonel Reed, who replied that heavy artillery fire was holding up the advance.

"When are you going to jump off?" Gerhardt shot back, "Get them started. Let me know when they jump off. And, make it soon."

He was even more furious when he later learned that Reed had not started the attack by the end of the day.

General Gerhardt next called for an all-out attack across the entire 29th Division front on July 13, believing the intensity of this attack would completely overwhelm the enemy. Again, the focus would be Martinville Ridge, where the Second and Third Battalions of the 175th, supported by two companies of the 747th Tank Battalion, would pass through the 116th and attack due west into Saint-Lô. The offensive would be supported by Army Air Corps fighter-bombers.

Gerhardt's plan might have worked if everything had gone as expected. What Reed knew all too well, and what Gerhardt had not yet learned, was that the tanks had not arrived due to shortages of fuel and ammunition. Additionally, the air assault was grounded by overcast weather conditions.

Under orders, the Third Battalion launched its attack but ran into an explosive wall. The Second Battalion joined the attack but without faring any better. Communication lines had been knocked out between battalion headquarters and some of their companies, leaving them deaf and mute in the melee.

A German Kettenkrad, an armored vehicle with motorcycle forks, brazenly roared down the road that followed the crest of Martinville Ridge, spraying machine gun fire at the passing hedgerows like a getaway car in a gangster film.

The attack of July 13 gained only a few hedgerows at a cost of 152 casualties. The 115th lost 108 men on the day with little ground gained. At that rate, the regiment would be entirely depleted before reaching Saint-Lô. This offensive was costing Gerhardt more men than were lost on Omaha Beach. The general's plan to seize the city from the east had failed, and now the 29th had no more room for subtle maneuvers. The only thing left for the division was to blast its way into town. In spite of the losses his troops were taking, Gerhardt suspected that the Germans were on the verge of defeat. Reports were that German prisoners displayed all the signs of a defeated army – exhaustion and demoralization. They had been in retreat since June 6 without adequate reinforcements. Signs of the effectiveness of artillery barrages on the German troops were everywhere, destroyed heavy weapons and equipment left laying in fields and along the roads. Gerhardt recognized that the men of the 29th needed a rest, so he gave them the day off on July 14. Rifle platoons were pulled from the front and brought back for a hot meal and a shower. The general took the opportunity to tour the front lines, paying particular attention to the amount of debris and animal corpses lying about. Dead and wounded men were removed from the battlefield as quickly as possible, but farm animals had been left where they lay and this was unbearable for the general. He ordered all of the dead animals to be buried immediately.

The offensive of July 15 was led by the 115th Regiment. Artillery stalled their early advances but by afternoon they had captured more German prisoners than they could count.

"Pour it on," Gerhardt exhorted Lieutenant Colonel Lou Smith, executive officer of the 115th. "This may be it!"

He told Colonel Reed, "We're going to keep at this now come hell or high water."

Later that afternoon, Gerhardt prepared a stirring message for his regimental and battalion commanders:

> The division has accomplished extremely good results today. The advance should be pressed at all costs. At 1930 hours, the full weight of all capabilities of the 29th will be launched to achieve our objectives prior to dark. Every individual in the division should lend his utmost to this end. Fix bayonets! 29, Let's Go!

On the ground, it was a different story. The men on the front lines had a difficult time understanding Gerhardt's wild optimism with the grim realities they faced on the battlefield.

The gains of the 115th on July 15 amounted to six hundred yards of ground at a cost of 119 casualties, with twenty dead.

In spite of his optimism, a bitterly disappointed Gerhardt called off the attack at nightfall, concluding, "We did all right today but did not make the grade."

Then, something startling happened. An hour before dusk, the German line on Martinville Ridge suddenly broke. The Second Battalion of the 116th, under the command of Major Sidney Bingham, pushed the Germans back to within a mile of Saint-Lô but found themselves isolated. Elated over the sudden progress, Gerhardt also worried about a German counteroffensive that could demolish the "lost battalion." In spite of his concerns, Gerhardt left them in place. Headquarters of XIX Corps passed down the order for a day of rest on the 116th. Bingham was ordered to hold tight and be prepared. The door they managed to crack open had slammed shut behind them.

Major Tom Howie assumed command of the Third Battalion of the 116th Regiment on July 13. Now, on the evening of the 16th, he was ordered to take his battalion, at half strength, into an attack on the town of La Madeleine the next day before dawn. There, he would join Bingham's Second Battalion and fight their way into Saint-Lô. The Germans had not launched a counter-attack and Bingham's men continued to fight as they held their ground. Howie was to take the most direct route to join Bingham, and then, together, into the town, while elements of the 115th and 175th cleaning out enemy pockets. The 175th had moved to control the Bayeux-Saint-Lô highway, relieving the First and Third Battalions of the 116th Regiment.

Howie's advance succeeded in reaching Bingham, and at seven-thirty that morning he was ordered to gather both battalions and move into Saint-Lô. Following a meeting with company commanders, the Germans laid down a mortar barrage on the command post. As Howie looked back to make certain his men had covered, a fragment struck him in the back, puncturing a lung. He died two minutes later.

Suffering intense mortar and artillery fire throughout the day, Company E of the 175th was left with fifty men and one officer, when its normal strength was one hundred eighty-seven men and six officers. The Germans were defending Saint-Lô with everything in their power but with the break in the German line, Saint-Lô was now on in sight. It remained that progress was measured in feet and yards, rather than miles and kilometers.

Driving into and walking through the ruins of Saint-Lô. (Maryland Military Museum)

Utilizing flanking actions around the German strong points, American troops entered Saint-Lô on Ollie's forty-eighth birthday - July 18, 1944. The victory was bittersweet as Saint-Lô was gained at a cost of nearly 11,000 Americans killed, wounded, and missing or captured in the Battle of the Hedgerows from July 7 through the 22nd. Of that number, the 29th Division suffered 3,706 losses. For this number of casualties, three to seven miles had been gained on the frontlines.[197] Ninety-five percent of the medieval city of Saint-Lô lay in ruins. The scene, reminiscent of the aftermath of the World War I battle of Flanders, was shocking to hardened warriors and replacements alike.

With Saint-Lô in American hands, General Gerhardt ordered the advance column to carry Howie's body into the town square atop a Jeep, leaving his flag-draped body in state on the rubble of steps leading to the ruined Sainte-Croix church in the city. A *New York Times* story immortalized the combat commander, headlined, "The Major of St. Lo," and the event inspired poet Joseph Auslander to pen the verse, "Incident at St. Lo."

Major Howie's body atop St-Lô rubble. (National Archives)

The day he entered Saint-Lô, Ollie took time to write to Mildred, unaware of his son's death two weeks earlier:

> Dearest: Haven't written for several days but not because I haven't thought of you many many times.
>
> The regiment has been very busy on the front and I am not much of a Regimental C.O. My nerves don't stand up as well as I had hoped – but guess I'll stand it. Am doubly glad that it will not be my job to take the 309 into action.
>
> This is quite a birthday and I am hoping that next birthday I will be with you, Buddy, Laura, Ollie III, Ted and his best.
>
> I am well, robust in fact and my spirits go up and down with the situation. Had a good long sleep last night so right now my spirits are up.
>
> Had a short V-mail from Buddy – he seemed happy. Bud must have been in Rome – hope he stays there.

Forty-three days after landing at Omaha Beach, the first major objective of the invasion, Saint-Lô, had been secured, but it was not time to slow down or stop. Commanders were already preparing for the next offensive, Operation Cobra, intended to achieve the long-sought breakthrough of the German lines. Once accomplished, the way would practically be open to Paris. After that, Berlin.

In the meantime, General Gerhardt kept his promise to his troops that once they reached Saint-Lô, they would have a well-deserved rest.

★★★★

Resting beside a Normandy hedgerow: peasants in the 19th century and an American MP in 1944.
(Library of Congress)   (Maryland Military Museum / National Archives)

29th Division Cemetery at La Cambe. (National Archives)

On July 23, the 29th Division cemetery was dedicated at La Cambe, about twenty-three miles north of Saint-Lô, close to Omaha Beach. General Gerhardt ordered that the grass be freshly mowed, each of the two thousand crosses and Stars of David whitewashed and in perfect alignment. The grounds were covered with American flags and the blue-and-grey of the 29th Division, along with regimental and battalion colors. A crude rectangular signpost at the entrance to the cemetery was topped with the division's blue and gray yin-and-yang symbol. The sign read, "This cemetery was established on 11 June 1944 by the 29th Infantry Division, United States Army, as a final resting place for officers and men of that Division who made the supreme sacrifice on the battlefields of Normandy. . . In command of this valiant legion of the Blue and Gray is Lt. Col. William T. Terry, Infantry, who was killed in action 17 July 1944." Terry was the commanding officer of the First Battalion, 175th Regiment, and the highest-ranking 29er to lose his life since D-Day.

It was a hot, humid day. Flags hung limply in the still air. The ninety-piece divisional band played mournful music.

General Gerhardt delivered a brief message: "Some of the men buried here were my personal friends. All of them, whether they believed in Christ, or Jehovah, or Allah, regardless of their creed, all of them are in the hands of a supreme being."

After the playing of "Taps" and "The Star Spangled Banner," Gerhardt returned to the podium. He had one more thing to say and wanted every 29er to join him: "Two-Nine, Let's Go!"

The next day, Ollie wrote to Mildred:

> Have been in command for one month and one day. Think little of myself as a Commander, and I pray I continue to be humble.
>
> Yesterday, the Division held the Dedication of a Division Cemetery. Poorly handled. Not impressive. Remember how shocked we were that they played gay music at the end of the first military funeral we attended? Its tradition, but I still don't like it.
>
> They played Beer Barrel Polka yesterday. Revolting.

During their days of rest and relaxation, officers were also given time off. On July 21, Ollie wrote to Mildred:

> Dearest: - Wrote you yesterday for the first time, I believe, in five days. Sorry about the lapse but I did think of you a lot. Last evening Lt. Col. Purnell, Ex., Major Beacham, Surg., Bounds (my driver) Ferris (my radio) and I all attended a movie in a little French town movie hall. Mickey Rooney in "Blonde Trouble." The picture was a good one for all of us and we all laughed a lot. Preceded by screen singing and it gradually warmed everybody up. Purnell is a Baltimore lawyer of very great ability and should actually be in command of the Regiment – he had earned it. After the show immediately to bed in a luxurious slit trench – wide enough and deep enough to have my cot in with six inches at the bottom. Bounds had broken out my slippers so that I took off my boots in my tent. My pajamas are at the Service Co., so I tried sleeping bare from the waist down – plus four blankets. You may be warm there but it is still cold here. This a.m. woke to a slow steady rain. Oh yes, the boys built my pup tent over the slit trench so it was absolutely dry and I even have a little shelf on each side to lay this on.

By the way you speak of going into a chapel to say a "little" prayer – next time try a "big" one.

★★★★

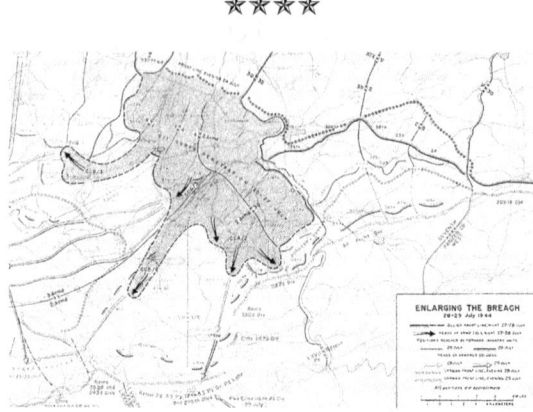

Operation Cobra launched on July 25. More than fifteen hundred B-17 and B-24 bombers flew overhead to their targets along the German front lines. Everyone, from political and military leaders to the men in the field, was feeling confident. Advances continued as thousands of men awaited their deployment to join the fight in Europe and the Pacific. Word circulated of the attempt on Hitler's life by his own officer corps. USO shows brightened spirits. The end, many thought, must be near. Americans could smell blood and the 29ers were ready to go in for the kill.

By July 28, the German lines west of Saint-Lô had been smashed wide open. The 29th crossed the Vire River and followed the Second Armored Division against an enemy on the verge of collapse. The 29th advanced toward the tiny village of Percy on a mission to ward off any possible German counterattack.

The next day, as the 29th had settled amid the hedgerows just north of Percy, General Gerhardt looked at his situation map with dismay. The front was littered with disjointed pieces of the haphazard American advance. There were gaps, hundreds of

Infantryman and tank in combat. (National Archives)

yards wide, that the Germans could easily exploit. They had moved in two Panzer divisions from the east, and Gerhardt feared they could easily break though and cut off supply lines from the coast. On the surface, enemy forces showed clear signs of a broken, retreating army.

In fact, they were anything but defeated. Fierce firefights were erupting across the entire frontline. Enemy tanks were suddenly prowling throughout the supposedly secure zone.

Colonel Reed's 175th was the spearhead of the 29th's attack in Operation Cobra. The regiment's adjutant, Captain James Hays, arrived at the crossroads hamlet of La Denisiere, roughly five miles north of Percy, at about 9:00 a.m. on July 29 with a small party of GIs. As he saw parts of the First and Second Battalions pass by, he knew that he was not where a regimental adjutant was supposed to be – on the front lines. Soon, he was receiving close-range fire from a German machine gun that killed two medics. The men took cover in a house at La Denisiere intersection. Hays looked out the window to see a German Mark IV tank in the crossroads firing away. Hays was captured but managed to escape the next day. Had his German captors searched him more thoroughly, they would have found a map outlining his division's positions in his pocket.

Breaks formed in the low overhead clouds as morning light appeared. Several positions had been hit in pre-dawn air raids, evacuating some twenty-three men for medical attention without any reported deaths. Elements of the Wehrmacht 2nd Panzer Division were approaching the front lines and artillery fire was increasing as the morning hours passed. The 29ers had seen few enemy armored vehicles in Normandy and were not prepared to deal with them now. Fire was coming in from the north as the Germans were attempting to establish positions behind the American lines and cut supply lines from Saint-Lô to the front. Colonel Reed entered his regimental command post near Villebaudon, a crossroads village about halfway between La Denisiere and Percy, under an increasing artillery barrage. Two infantry companies of the 60th Panzer Grenadier Division imperiled the entire 175th front. While seven tanks of the 2nd Panzer Division were tied down in a street battle in Tessy, the 116th Panzer Division, the Greyhound Division, appeared out of nowhere. A Second

Battalion captain saw the muzzle flash of a German 88 through his field glasses and yelled, "Hit the dirt!" saving men's lives.

In *If You Survive*, George Wilson wrote about the July 30 battle:

> Exploding shells flashed everywhere, raising much dust and smoke. In wild panic, the men dodged about screaming and headed for the rear. Their eyes were wild with fright and tears streamed down their contorted faces. They were in complete panic.[198]

That morning, German tanks and infantry of the 60th Panzer Grenadier Division overran the Second Battalion of the 175th Regiment, cutting communications with their headquarters. Their last reports had the battalion moving to high ground but no one knew exactly where.

The Second Battalion pulled back, leaving their aid station, located in a barn just outside of Villebaudon, with its unarmed men, open to the Germans. Chaplain for the 175th Regiment, Reverend Gerard Taggart, emerged from the barn with hands raised, opening doors clearly marked with a red cross, shouting "Nicht schiessen!" (Don't shoot!) "Wounded!" To convince the Germans that they posed no threat, he brought bloodied and bandaged men out of the barn, pleading for safe passage. After Taggart and the wounded advanced about ten yards, the Germans opened fire, killing ten men in cold blood. Taggart fell to the ground and feigned death.[199] He arose to return to the barn to administer aid to the wounded. Corpsman Don McKee, having just returned from replenishing aid kits, also laid on the ground pretending to be dead. A German boot stepped right next to his head. Lieutenant Colonel Millard G. Bowen and Captain Lester Maize were among the wounded murdered in the aid station.[200] Taggart was later awarded the Silver Star for his actions that day. The citation concluded:

Father Gerard W. Taggart

Lester Maize (Courtesy Barbara Sall)

"The courage and devotion to duty displayed by Chaplain Taggart reflect great credit upon himself and the Military Service."[201]

McKee later recalled in a letter to his fellow medics, "The men knew they were facing a different German that day. They were not the young, inexperienced or non-German soldiers they had taken prisoner near the beaches of Normandy. They were members of the Panzer units that had overtaken most of Europe in swift, effective methods."[202]

The men of the 60th Panzer Grenadier Division had been fighting since the invasion of Poland in September 1939. From there, they participated in the invasion of France and returned to the Eastern Front to fight in the siege of Leningrad. Battered, they picked up stray troops as they marched back to France.

Staff officers no longer knew where the front lines stood. Communications with forward positions had been knocked out. General Gerhardt ordered Reed to inspect the line and report back: "Ollie, you've got to get up there and find out what the hell is going on."

At 12:46, Colonel Reed, his driver, Sergeant Vaughn Bounds, and Lieutenant Curtis Fitzgerald evaded artillery and small arms fire as they sped toward the First Battalion headquarters in an orchard near the crossroads in Villebaudon.

(openstreetmap.org)

At one o'clock, German artillery pinpointed the 175th Regimental Command Post, bringing down a severe barrage that scored a direct hit on the headquarters. Captain Clarence Major was seated next to Lieutenant Colonel Edward Gill when they heard a shell coming. Everyone hit the ground, Captain Major fell on top of Lieutenant Colonel Gill. A shrapnel fragment pierced Major's thigh and passed into Gill's chest, killing him instantly. Captain Opie E. Chick was also killed. Only two men emerged uninjured. Another shell hit the S-1 tent killing Lieutenant John W. Chittick.

Clarence Major
(Courtesy Rebecca Robbins)

John Chittick
(Courtesy Tom Bedecarre)

An estimated two squadrons of German infantry, supported by a Panzer tank, got within a hedgerow of the command post but were repelled by Sergeant Abe Sherman, a forty-seven-year-old World War I veteran and native of Baltimore, Maryland. Sherman organized every remaining man – cooks, couriers, secretaries – men who hadn't fired a gun since basic training, handing out weapons and ammunition, urging the men to defend their position. With complete disregard for his own safety, Sergeant Sherman organized and managed a successful defense of the regimental headquarters, for which he was awarded the Silver Star.

Arriving at the First Battalion headquarters, Colonel Reed,

Vaughn Bounds
(Courtesy Joe Bounds)

Lieutenant Fitzgerald, and Sergeant Bounds were wounded by shrapnel from an artillery round as soon as they stepped from their Jeep. Fitzgerald and Bounds, in spite of their wounds, managed to get Colonel Reed into the vehicle and to the closest medical clearing station. Bounds drove the Jeep, while Fitzgerald sat in the back with the severely wounded Colonel Reed, literally holding his abdomen in place. Colonel Reed was conscious when they arrived at the clearing station near the 224th Field Artillery Battery position, asking Sergeant Bounds, over and over, not to leave him.

Curtis Fitzgerald
(Courtesy Barbara Fitzgerald)

Lieutenant Odis B. Summers in the headquarters of the 224th was the first to report that Colonel Reed had been seriously hit to Captain Thomas Neal in the Regimental headquarters of the 175th.

Colonel Reed died soon after from the loss of blood.

★★★★

At 1:35 p.m., General Gerhardt issued an order to the 175th command, telling them, "do not, repeat NOT, abandon," their position. At 2:00 p.m., he issued a regiment-wide message that the situation was under control, closing with, "Two-Nine, Let's Go!"

At 3:24 p.m., Lieutenant Colonel Purnell was placed in command of the 175th. "See that he gets promoted," Gerhardt ordered.

In four days, from July 29 to August 1, 1944, the 29th suffered nearly seventeen hundred casualties. Within days, the German lines and resolve cracked, and the Allied breakthrough was accomplished.

It was said that had Reed and the others survived July 30, the bloodiest day in the war for the 175th Regiment, they stood a good chance of surviving the war.

On August 1, Colonel Reed was laid to rest in the 29th Infantry Cemetery at La Cambe between two privates – twenty-four-year-old Robert J. Anderson of Camden, New Jersey, and Clyde T. Angel, 27, from Elizabethton, Tennessee. For the burial ceremony, General Gerhardt had the sign at the entrance changed to give special recognition to Colonel Ollie W. Reed, the highest-ranking member of the 29th Division killed in combat in World War II.

The 29th went on to halt the German counter-offensive, enabling the Allies to maintain their march through Normandy toward the Rhine. The division crossed into Germany in February 1945.[203]

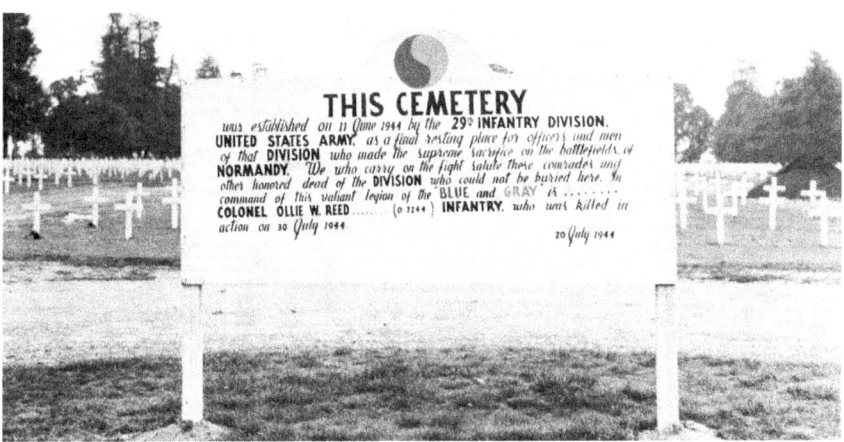

Entrance to the 29th Division Cemetery in La Cambe, France, August 1944, with Colonel Reed's name. (National Archives)

Above, a map from *29 Let's Go!*, a booklet produced by Stars & Stripes. (Stars & Stripes)

Three postcards produced by the 78th Division as they fought their way through Europe. (Reed family collection)

Love and Sacrifice 283

CHAPTER TWENTY-TWO

Love and Loss

Ollie, Mildred, and Buddy, 1919

Love is not love
Which alters when it alteration finds,
Or bends with the remover to remove.
O no, it is an ever-fixed mark
That looks on tempests and is never shaken;
It is the star to every wand'ring bark,
Whose worth's unknown, although his height be taken.
 - William Shakespeare, Sonnet 116

JULY 1944 WAS AN UNUSUALLY RAINY MONTH in Manhattan, Kansas, with nearly seven inches falling. Throughout the summer, temperatures in Manhattan had been well above one hundred. When August arrived, the rains suddenly stopped and the daytime mercury dipped to the ninety-nine mark. Afternoon clouds often brought an evening shower to the river valley town, clearing the hazy air of the day.

A slight breeze blew through the trees outside 1111 Bluemont Avenue, the temporary home of Mildred, Laura, and six-month-old Ollie III, on the evening of August 13. A knock at the door made mothers and wives of service members flinch fearing the worst. Laura had already received word that her husband was missing in action, but she held out hope for the best.

This evening, the knock on the door brought a Western Union telegram addressed to Laura:

> THE SECRETARY OF WAR DESIRES ME TO EXPRESS HIS DEEP REGRET THAT YOUR HUSBAND LT. OLLIE W. REED, JR. WAS KILLED IN ACTION ON SIX JULY IN THE EUROPEAN AREA.
>
> J.A. ULIO THE ADJ GENERAL

Laura fell to pieces. Mildred was shaken but had to remain strong for Laura and the baby. This was the moment her son had warned her before he left for war.

An hour later, as the family grieved for Bud, there was another knock on the door with news of Ollie's death.

Mildred later wrote of a premonition Laura had in her sleep around the time of Buddy's death:

> Laura had dreamed the night of July 5 (there's a time difference, you know) that Buddy had come to tell her good-bye.
>
> These were hard blows to take but Laura had to be brave for me and I had to be brave for her. Ted helped keep us on an even keel. A college man must be fed and little Ollie absorbed us with his daily needs. Laura worked as a nurse's aid at the hospital. So the numb days slowly passed.

The local newspaper, the *Manhattan Republic*, reported the tragedy:

> August 17, 1944
>
> Tragedy stuck twice for the Ollie William Reed family 1111 Bluemont, Sunday evening, when they were notified that Lt. Ollie Reed had been killed in Italy, July 6, and a short time later that his father, Col. Ollie Reed, had been killed in France on July 30.

In her steadfast humility, Mildred seldom spoke of this tragic day.

In her memoirs, written for her grandchildren, this was all Mildred had to say about her own grief and emotions:

> The following incident is not to elicit sympathy – but no record

of my life would be complete without mention of this dark day.

On August 13, 1944, a telegram came bringing news of my son Buddy's death in Italy on July 6. An hour later, there was another 'War Department regrets' message. My husband had been killed in France on July 30. I don't like to talk about this unless I tell of how the Lord comforted me. Before the men had gone overseas, I bought five little daily reading books called "Strength for Service." They were for Ollie, Buddy, Bud's wife Laura, Ted and me. The men wrote that mail was sporadic but it made them feel closer to us to know they were reading the same scripture and comments each day that we were. The next day after our devastating news, I said, 'I wonder what the last verse was that Ollie read?' I looked up July 30 and read, 'All things work together for good for them that love the Lord.' That verse had proven true for us so many times through the years that we claimed it as our special family verse. Wasn't that wonderful of the Lord to assure me that it was better to have my loved ones safe with Him than suffering further ravages of war?

General Gerhardt took time the day Ollie was killed to write a letter to Mildred but the army had an outdated address on file, so it was not received until some time later.

July 30, 1944

Dear Mrs. Reed:

I wanted to write to you and tell you about Ollie's death and his service with the regiment of this Division since he joined us north of St. Lo some time back. We were all delighted to hear that he was being assigned to the Division and he was given command of a regiment right off. He was extremely active and worked himself into the situation without any delay. Had lots of good ideas and really was on the ball. Recently I had the honor of awarding him the Combat Infantryman's Badge. Just prior to that we had a shooting match with pistol and carbine just to be sure he could shoot – and he certainly could.

We jumped off three days ago after a week's rest, and his regiment led off in the move, and pushed down vigorously and finally took up position where we have been receiving considerable trouble from the enemy. He was constantly out with his battalions and was severely wounded this morning as he was coming back from a visit to one. He died this afternoon in our Clearing Station, and, of course, we all feel very badly about it. I am sure that he was happy in his last assignment, and you can be proud of the way he wound up his career.

Several days ago we were discussing days at Leavenworth in 31 and 33, and he mentioned the time you all were in a play at Benning with some of our mutual acquaintances in those parts.

With kindest regards and deepest sympathy, I am,

Sincerely Yours,

C.H. Gerhardt
Major General, U.S. Army,
Commanding

At the bottom of the page, Gerhardt added a handwritten note:

I hope the son in Powder River is OK. C.

Lieutenant Colonel William Purnell, Colonel Reed's executive officer, also wrote to Mildred from the battlefield:

Sept. 4

I have just received your letter saying you have received notice of Colonel Reed's death. I hope it will bring some measure of comfort into your sorrow to know that he met the end of a very fine and gallant soldier. During a particularly severe fight when communication with the forward troops had been interrupted, Colonel Reed drove up to the front line to get what every commander must have - accurate knowledge of what is going on. While there, both he and his driver were struck by fragments of artillery. He did his duty and was killed in the midst of doing so.

Major General William C. Purnell
(Maryland Military Museum)

That is the way we all want to go should our turn come around.

Colonel Reed had only been with us five weeks, but as his executive officer, it was my good fortune to be closely associated with him. I enjoyed so much his warm friendliness and deep and constant concern for his men. Their welfare was always his first thought and I admired that very much.

Only a few days before the action in which he was killed, he and I and one or two others of the staff had a most enjoyable dinner, on the first real chicken we had seen in many, many weeks. It was the fruit of Colonel Reed's persuasive manner with a French farmer near our command Post who prepared it for him.

I can assure you that Colonel Reed was in fine spirits right up to the last. He spoke almost continually of you and seemed delighted as a young boy when your delayed mail finally came through.

The 309th Infantry Regiment held a memorial service in the regimental chapel at Camp Pickett on August 30, 1944. Many tributes were paid in Colonel Reed's memory. Among them was one recorded without attribution:

309th Regiment memorial service at Camp Butner, North Carolina.

> He was deliberate in his thought and planning; strong and thorough in execution; he carried out the fundamentals of his beliefs in every day life. He took personal interest in his men and knew the majority of them by name.

General John K. Rice, who led the 78th Division in Europe and whose son had been killed in Italy near Anzio, wrote about Colonel Reed, a man whose career paralleled his from the very beginning:

"He had a warmth of heart that produced in him a continuous deep personal interest in his associates, whether they be soldier, officer, relative or friend."

The newspaper of the 78th Division, *Parade Rest*, published a series of tributes in August 1944:

To the Lightning Division Colonel Ollie W. Reed will always be remembered as one of its first members and finest characters. He was a superior officer who was always busy on behalf of the 309th Infantry and the Lightning Division. He was an outstanding leader…the type that the last man in his regiment would follow to the end.

Colonel Reed had a heart of gold and was ever doing something for someone else. He was the type of soldier that members of the 309th Infantry, and of this Division, should set up as their model and guide.

The commanding officer of the 309th who succeeded Colonel Reed, Lieutenant Colonel John G. Ondrick said of Reed:

> With the death of Colonel O.W. Reed I feel that the U.S. Army has lost an exemplary soldier. It was with particular regret and sorrow that we of the 309th received the news of his death. Colonel Reed was responsible, more than anyone else, for the splendid training which we now enjoy. We of the 309th Infantry regret his passing.

Commanding officer of the 78th Artillery in France, General Frank Camm, wrote to the newspaper:

> A great soldier has passed, a soldier's soldier, a man I would like to follow into battle. He would look out for me even more than he would expect to look out for himself. His unselfish attention to duty and understanding of his fellow soldiers contributed immensely to the successful record of the Division. Colonel Reed is gone, but his fine influence will live forever in the hearts of his comrades.

Even the editor of *Parade Rest*, Sergeant S.P. Garramore, wrote a personal letter to Mildred along with copies of the tributes.

> I want you to know that this edition was not the idea of a single individual…rather it was the spontaneous request of the men of the 309th. In my 17 years of newspaper experience I have never

seen such spontaneous reaction on the part of a group of individuals…and it does make me proud.

Garramore wrote of speaking with his colonel before Ollie left for his new assignment:

> You perhaps recall my speaking to the Colonel shortly before he left the 309th. At that time I told him that his men wanted to say goodbye to him and that they would be hurt unless the request was granted. They are still the same men, Mrs. Reed, and they still love "The Old Man" as they called him and I feel sure that you will take no offense when I use the term "Old Man" because I can assure you it is used with admiration, reverence and respect. I can only say…We knew Colonel Reed and because we knew him and loved him, we understand.

Page from Mildred's scrapbook about the 309th

Ollie Reed was, indeed, a kind and caring leader. Every one of the soldiers under his command was like a son to him. But, are these the ideal attributes for a battlefield commander? As his 1933 evaluation at the Command and General Staff School concluded – Captain Reed was qualified for General Staff, as well as service at the division and corps levels, but the one box marked not qualified was in Command.

As fellow Kansan, Major General Clarence Huebner, said about his strict command demeanor, one for which he was disliked by men serving under him: "When you take over a command, you start off being an SOB and later become the good guy, but you can never start off being a good guy and later become an SOB!"[204]

Every month, a copy of *Reader's Digest* arrived in the mail addressed to Laura. Buddy provided a lifetime subscription for Laura so she would receive a little present from him every week.

Headstones for Ollie and Bud in the Normandy American Military Cemetery in France. (Robert Uth)

On February 18, 1949, Colonel Ollie W. Reed and First Lieutenant Ollie W. Reed, Jr. were reunited in the Normandy American Cemetery at St. Laurent-sur-Mer, overlooking the English Channel from the bluff above Omaha Beach.

Mildred and Laura were informed of the reburials with full military honors by separate letters dated May 17, 1949. Mildred had hesitated asking the government for their placement side-by-side but, as she wrote to Major General Feldman of the Quartermaster Corps in gratitude, "it had brought an immeasurable sense of peace to my lonely heart to know that they lie beside each other; that even in a foreign land they won't be lonely now."

They are the only father and son among the 9,385 who are buried in the Normandy cemetery. The inscription on Bud's cross incorrectly gives his birth state as Missouri, he was born in Connecticut, and the date of his death. In the confusion of the first days of combat, Lt. Reed's death was incorrectly recorded as July 6. He was killed on July 5, 1944.

A simple bronze tablet above the baptismal font in the Fort Leavenworth Post Chapel where Buddy and Ted were baptized in 1931, was dedicated on September 27, 1947, in honor of father and son Reed. Mildred explained the comfort of having a memorial on U.S. soil. In writing to Chaplain Major Arthur H. Marsh in 1947 about the placement of a plaque in the chapel, Mildred explained: "Being an army family we've had no home, so I do not plan to ask for their bodies to be brought back to the U.S. for burial, but I would like to feel that their memory was honored in someplace they loved."

Laura had a difficult time coming to grips with Bud's death, believing that one day he would walk through the door. In 1945, she went ahead with her flying lessons, perhaps to soar closer to her beloved. Although devastated over the death of her young husband and the father of her infant son, she followed Buddy's advice and remarried. On August 19, 1945, she wed cavalry officer and West Point graduate, Lieutenant William Vaughan, in the post chapel at Fort Riley, Kansas. Bill adopted Ollie III as his own son and, in turn, Ollie adopted his step-father's name. They started married life much the same as Ollie and Mildred, serving with the Army of Occupation in Germany and Japan. Bill and Laura lived a long, happy life together. He continued his education and became a successful construction engineer.

Laura with the plane she was learning to fly.

Laura and Lt. Bill Vaughan wedding August 19, 1945

After Bud's death, Laura finally learned the true story of her real mother from her Aunt Helen. With the knowledge about her birth mother and her work in education, Laura dedicated her life to completing her education and setting upon a career as a kindergarten teacher. Laura graduated from college the same day her son graduated from high school. In 1979, Laura had her father's remains moved to the Allegheny Cemetery in Pittsburgh to rest beside her mother, Laura Gillmore Sloman, and placed a headstone with both of their names on the previously unmarked grave. Laura passed away in 1999, never forgetting her first love.

Mildred, left, at the Casa de Alfabetizacion in Mexico.

After her husband's death, Mildred traveled ceaselessly across the United States, visiting friends and relatives, and to rural Mexico to volunteer in a literacy program. She never owned another home after the "tucco" house in Merchantville, New Jersey. The road was her home, as it had been with Ollie through so many years. Her address book was the size of a telephone book. If ever a friend or relative were in need, Mildred would appear on their doorstep to help.

Mildred never remarried, but in a 1980 letter to her friend Esther in Kuna, Idaho, she recounted a rekindled courtship and close call:

> You have asked me if I ever considered re-marrying. You know I always talk things over with the Lord and this is one time His answer was loud and clear.
>
> A year after Pop was killed, a man I had known and liked in high school came back into my life. He had graduated a couple of years ahead of me. He had gone to Argentina and had written me fascinating letters. I remember, after reading one of his letters aloud to my mama, I said, "Oh, I wish Peter loved me. It would be such fun to be married to Peter." My mama said, "I don't think it would work. You both are dreamers. If Ollie develops into the man he shows promise of being, I'm sure you'd be happier with him."
>
> Peter took me dining and dancing — things I never expected to do again. (I was 48 the summer Pop died.) We had pleasant times together. After awhile marriage was mentioned, but we put off serious discussion until spring.
>
> Spring came. The night before Peter was to come from Omaha, I prayed, "Lord, I haven't consulted you about this, but I want your approval of my marriage to Peter. If there's any reason you don't want me to marry him, let me know; but, please don't search too hard for objections, because I think we'd be congenial companions, and I want to marry Peter."

> I hadn't much more than fallen asleep, when the phone rang. It was Thomas Jefferson <sic> Duvall - the preacher my mama had married in 1915, and with whom she hadn't been happy. He married very soon, a spinster who was wealthy and who took good care of him. I hadn't any loving feelings about him — in fact, I didn't know if he were dead or alive — and didn't really care, good Christian that I am!
>
> Mr. Duvall told me he had heard that I lived in Manhattan. He was changing buses and had a three-hour wait. Would I come to the bus depot to see him?
>
> Of course, I went. I brought him home and fed him. He was a frail little old fellow. I forgave him that he hadn't been nicer to my mama. He said he had loved my mother and that I was like her - sweet-talkin' southern gent to the end! I put him on his bus and didn't hear another word from, or about, him until six years later; his wife sent word of his death to the Norton church.

Reverend Duvall and his wife, Cora, had been living in California after his posting in an Abilene, Kansas Baptist church. He retired from the First Baptist Church in Santa Paula, California and lived out his days on Greenleaf Street in Los Angeles.

For Mildred and Peter, fate would not hold them together.

> The future had looked ideal for my mother, but it didn't turn out that way. I expected my future with Peter would be rosy. I relived my mother's experience and knew that the Lord had sent me my disappointing answer.

In 1958, Mildred visited a non-denominational religious center in Stevenson, Maryland, just outside of Baltimore, called Koinonia.[205] She was interested in learning how to teach reading to illiterate adults. This training took her to the Casa de Alfabetizacion (House of Literacy) in Valle de Bravo, Mexico, in the mountains about one hundred miles southwest of Mexico City. The casa was a one hundred-year-old hacienda in the center of town with about twenty-five rooms that were used for classrooms and living quarters.

Although not fluent in Spanish, Mildred spent years, off and on, at the casa, establishing and running the recreation center that she and her many friends equipped. Across the bottom of the sign welcoming visitors it read, "All may learn to read and write here without charge." The casa employed the Laubach method of teaching, "each one teach one," that was developed by Christian missionary Frank Laubach in the Philippines during the 1930s, where Mildred may have first come into contact with Laubach and his literacy work. The program began with fifty-seven students and quickly grew to 140, ages nine to thirty-five. Mildred would draw prospective students right off the streets and into her game room to introduce them to the literacy program. In Mexico, Mildred returned to her childhood nickname of Mit when locals had difficulties with Mildred. She was happy with "Tia Mit," Aunt Mit.

In 1962, Mildred's Uncle Lealon Ward, then quite ill, asked Mildred to come to Kuna, Idaho, to look after his wife, Myra Sarvis Ward. After several months, Mildred's Auntie, as she called her, told her, "Go back to your Mexicans where your heart is."

She returned to Kuna some time later when Auntie broke her ankle. While she was away, Roland Hosea, co-founder of the Casa de Alfabetizacion, and his wife were killed in an automobile accident. Mildred was deeply sadden by Hosea's death, a man she loved as a son. With his loss, the program began to decline. After suffering a heart attack in 1972, Mildred knew she could no longer live in Valle de Bravo, so she returned to Mexico for a final time in 1973, hitching a ride back to Kuna with her friend Pamela Brewer's niece, Janie.

> One day, Pam's niece, Janie, an odd girl who made pottery and did leather tooling, came into the room where I was working. She said, "Would you like to have me take you to Kuna?" I was astonished. "Why would you want to go to a podunk place like Kuna?" She said, "I have some business to check on in Tucson, and could take you on up there." I was speechless with delight. That meant: I could take my wicker rocker, hamper, stools, baskets.
>
> In exchange for the ride, Mildred promised to read books aloud during the entire drive. Over the course of twenty-three hundred miles, Mildred read five books aloud, including *Medical Missionary* and *Lady Llama* about Tibet.[206]

Mildred lived the life of a vagabond, frequently visiting her son Ted and his wife, Elizabeth, in Oregon or Washington, DC where she spun magical stories of life in a different time for Ted's daughter, Maryalyce, who loved spending hours listening to her grandmother's stories of the ages, just as Mildred's grandmother had done with her. When Maryalyce asked about the tragedy brought to her grandmother's life in the form of two telegrams with word of her loved ones' deaths, she quoted John Bailie's *A Diary of Private Prayer*, "I bless Thee for Thy hand upon my life, and for the sure knowledge that however I may falter and fail, yet underneath are Thine everlasting arms."

Mildred also regularly dropped in on Laura and Bill Vaughan, as they moved around the country, from Oregon to New York to Ohio, especially to see her grandson, Ollie III. When Mildred was present at the time of a birthday, she would always bake a cake and decorate it for her three Ollies. By the time Ollie III reached middle school, he preferred to be called Bill. Ollie was the name of a puppet, as in, *Kukla, Fran and Ollie*, and he bristled at the fun other kids poked at his name. His grandmother approved of his name change from Ollie to Bill.

Ollie III and the birthday cake for the three Ollies.

Sometimes Mildred could overstay her time, especially when she took charge of the household, but her grandchildren and great-grandchildren loved Mildred dearly. The memory of the woman they called Buppie remains vivid in their minds.

Among her many stopovers, Mildred paid a 1980 visit to Betty Sarvis in Indiana. Betty's life had been tough as she never quite recovered from Bud's death, suffering a nervous breakdown upon hearing the news. She was hospitalized, her husband abandoned her, and her children were placed in an orphanage for a time. She smoked heavily and enjoyed beer, but Mildred made no judgments, simply offering love and compassion. After her visit, Mildred penned a letter to Betty, recounting the story of a friend's nervous breakdown, telling Betty, "Life is made up of sunshine and shadow – and my life is no exception. I was brought up on the saying, 'It isn't so important what happens to you, it's how you take it. Love from your Mommie-Mit."

Nancy Campbell graduated from William Jewell College with a degree in French, aspiring to become a teacher, becoming a bookkeeper and personnel manager instead. She was briefly married to James Fletcher Dicken, a handyman, in 1950. Mildred attended Nancy's wedding, weeping as the couple walked down the aisle, envisioning the groom was her Buddy. Nancy died, alone, in the Kansas City suburb of Fairway in 2004 at the age of 83.

Looking back, Mildred recorded in her memoir:

> We take what Life deals us and don't whine. Ollie and I realized that sometime we would pay for the happiness we had had – and since I am the one left alone, I can take it. When his overseas orders came, I said, "What if you don't come back?" I can still see his love-filled eyes shining at me as he said, "You'll get along." So, I have to because he expected me to.

Mildred's friend, Betty Sweet, sent a poem she thought captured her friend's life story:

> I've bartered my sheets for a starlit bed,
>
> I've changed my meat for a crust of bread.
>
> I've traded my book for a sapling cane.
>
> And I'm off to the end of the world again.[207]

CHAPTER TWENTY-THREE

Epilogue

Ted Reed, as director of the National Zoo
in Washington, DC, with a rhinoceros. Undated.

TED REED GRADUATED from Kansas State University and married his college sweetheart, Elizabeth Crandall, the editor of the college yearbook, soon after graduation. At the age of thirty-four, Ted became the director of the National Zoo in Washington, DC, and is credited with making it one of the world's leading zoological parks. When he took charge of the park in 1956, Ted described it as a "zoological slum."

As Ted was steadily making improvements to the zoo, tragedy struck two years after he had taken charge of the National Zoo.

On May 16, 1958, two-and-a-half-year-old Julia Ann Vogt from British Columbia was mauled by two lions, Pasha and Princess, after slipping through a guardrail. Questions abounded as to how the child could have gotten through the rails, but calls rang out for the lions to be killed. Director Reed refused to euthanize the lions, focusing instead upon the improvements to make the zoo safe and enjoyable for visitors.

Ted managed to turn the tragedy into a call to action for further improvements at the zoo, convening a commission to review safety issues. He pioneered the transformation of zoos into research facilities, developing the Conservation Biology Institute in Front Royal, Virginia, also part of the Smithsonian Institution. He brought the first white tiger to the United States from India in 1960 and is probably best-known for bringing the giant pandas, Hsing-Hsing and Ling-Ling, to the National Zoo in 1972 as part of the opening of diplomatic relations between the United States and China. Ted ran the National Zoo as if it was his command, like his father – firm but fair. He demanded excellence and imagination, while relying on the knowledge and professionalism of each member of the staff to do the best job possible. He cared about his people and they, in turn, cared about him.

Elizabeth Crandall and Ted Reed wedding April 20, 1945.

Ted and Elizabeth spent their honeymoon at the Kansas City Zoo. "We had a good time, no matter what she says," Ted told the *Washington Post*. Elizabeth, besides raising two children, also was a surrogate mother to many of the zoo babies throughout the years. Their children were used to having exotic animals living in the house, turning their home into a very popular place to hang out for their friends. Ted may have managed the National Zoo in his father's style of a colonel but he made a point of raising his children with love, affection, and fun, without military-style hierarchy.

Ted and Elizabeth's son, Mark, followed in his father's footsteps, graduating from Kansas State University in Zoology and is today the executive director of the Sedgwick County Zoo in Wichita, Kansas. Mark's sister, Maryalyce, sought adventure in faraway places, inspired, no doubt, by her grandmother's stories, and is now living in a remote region of New Zealand. Maryalyce's daughter, hearing stories of the exciting zoo life her mother lived, spent part of her college years observing lowland gorillas in Africa.

Ted's wife, Elizabeth, died of cancer in 1978 and, in 1980, Ted married Dr. Sandra Foote. Ted passed away in Milford, Delaware, on July 2, 2013.

Bill Vaughan, born Ollie W. Reed, III, looked back at the events of his earliest years - the deaths of his father and grandfather, and the news of their deaths arriving the same day to his mother and grandmother - as the defining moments of his life. He felt driven toward public service as his calling and had planned to attend West Point, just like his father, but just as it almost knocked his father out of the academy, he was refused entry due to color blindness. Bill worked on Capitol Hill for 36 years and then for the Consumers Union, before retiring.

Bill Vaughan and Luther Davis, the artist who painted Colonel Reed's portrait in 1944, at the World War II Memorial in Washington, DC. (Dennis Whitehead)

"All my life there has been a melancholy tune, where I often think, I've lived for X percent longer than my father, and am I really doing things that would make him proud?"

In 1980, at the age of eight-four, Mildred sat still long enough to gather her thoughts and old letters in her self-published memoirs, *Letters from Mit*, written for the benefit of her grandchildren. She closed the chapter about her beloved husband with a stanza from the poem "There is No Death," by an anonymous author:

> You call it death – this seeming endless sleep:
>
> We call it birth – the soul at last set free.
>
> Tis hampered not by time or space.
>
> You weep. Why weep at Death? Tis Immortality.

Another poem concluded the chapter about her son, possibly penned by Mildred herself:

> Time cannot dim the cherished thoughts
> of loved ones who have gone,
>
> Nor age those brave and gallant souls
> whose memories linger on;
>
> They live enshrined within our hearts, forever young and gay,
>
> Their dreams are ours to cherish still,
>
> Their hopes will light our way.

Mildred never dwelled upon her grief and tragic losses. Rather, she always had faith that God would look out for her and that whatever had occurred in her life was God's will, and she abided by it.

Mildred's thoughts went out to the young men she and Ollie had known through the years who lost their lives in the war, particularly those from Wentworth and West Point, as well as members of the 309th Regiment, those left behind in the Philippines, and their families:

Hector Polla (USMA)

Hector Polla, the son of Italian immigrants from Higginsville, Missouri, whose Italian-immigrant parents worked very hard to send him to Wentworth and then West Point, entered the academy with Bud Reed and graduated in the class of 1941. He was captured by the Japanese in Bataan, survived the Bataan death march to the Cabanatuan POW camp, where he was transferred two days before its liberation to an unmarked ship that was sunk by Allied bombers, along with 1,600 others.

Kenneth Griffiths (USMA)

Kenneth Griffiths, a Kansas native and Wentworth graduate, and his Lexington wife, Mildred's old neighbor, Sarah Ann Aull, sailed to the Philippines with the Reeds on the *U.S. Grant*. Mildred and Ann were close friends throughout their time in Manila. Lieutenant Griffiths, stationed with the Philippine Scouts, remained behind to fight and was captured in the battle of Bataan. He died a prisoner in the Cabanatuan POW camp. He had contracted cerebral malaria but no drugs were available for treatment in captivity. A daughter, Ann Quarles Griffiths, was born on December 3, 1941. News about the birth of his daughter reached him in the Philippines a week after war had been declared.

Sandy Nininger (USMA)

Sandy Nininger entered West Point at the same time as Bud, graduating in the class of 1941, and was the first USMA graduate to die in World War II. He was killed in the fighting to defend Bataan. Wounded several times and out of ammunition, Nininger killed three Japanese soldiers in hand-to-hand combat before succumbing to the loss of blood.

Lieutenant Colonel William W. Murphey, whose wife, Mae, was Mildred's good friend and with whom she evacuated the Philippines, was captured at Bataan and survived imprisonment until 1945 when he died in the Japanese POW camp at Honshu.

Ted Nankivell, West Point class of 1941 and Wentworth student, who tried so hard to help Bud with math, joined the Army Air Corps, and was killed on August 8, 1942, when his bomber developed engine trouble on a landing approach.

Ted Nankivell (USMA)

Captain Robert Sweet (Courtesy Lt. Col. Bob Sweet, USAF ret.)

Captain Robert E. Sweet was Bud's good friend from childhood throughout their short lives. Bob Sweet died on August 19, 1944 when an equipment malfunction brought his B-24E down in a training exercise over the Nevada dessert. Prior to his death, on July 29, 1944, a ceremony awarding Sweet the Distinguished Flying Cross for his meritorious actions helping to rescue the crew of a crashed B-24 was postponed so he could attend the birth of his twin sons that day. The medal was posthumously presented to his wife, Helen, at her husband's burial in Arlington National Cemetery a month later. Bob Sweet's sister, Anne, and her husband, Ray Ortlund, nicknamed their son Bud in honor of Bud Reed.

It was these young men, among many more, and their families, known and unknown who had been impacted by the war, for whom Mildred grieved long before she ever gave a second thought to her own losses and pain.

Besides Ollie and Bud, only one other father-son pair was killed in World War II. Machinist's Mate Thomas Augusta Free and his son, Seaman William Thomas Free, were killed aboard the USS *Arizona* when the Japanese bombed Pearl Harbor. They rest at the bottom of Pearl Harbor. Also on the *Arizona* were seventy-nine sets of brothers, sixty-three of whom died in the attack. Among those were three groups of three brothers - only one surviving from each.

Perhaps the best-known familial loss during World War II were the

Joseph, Francis, Albert, Madison, and George Sullivan

five Sullivan brothers, who lost their lives aboard the USS *Juneau* when the light cruiser sank in the Pacific Ocean during the battle of Guadalcanal in November 1942.

Five Borgstrom brothers from Utah enlisted and after four of the brothers were killed, the fifth brother was found and returned home.

The Niland brothers, two of whom are buried in Normandy, were the inspiration for the film *Saving Private Ryan*. The four brothers went to war, and when three were believed to have been killed (one was a POW), the fourth was returned home.

From these cases, the Sole Survivor Policy was enacted in 1948 preventing a family member from being sent into combat if another family member had been killed in combat.

With age, Mildred's travels slowed. She continued to use Kuna, Idaho for her base, where she compiled her memoirs, *Letters from Mit*, using a childhood nickname originating with a cousin who could not pronounce Mildred.

Ted had been living on the Eastern Shore of Delaware since retiring from the National Zoo in 1984. After suffering another heart attack, Mildred decided to join him in on the Eastern Shore, where she and Ollie had talked about retiring, and a place Ollie dreamed of owning a thousand acres while he was overseas.

Though Mildred's memories fell victim to Alzheimer's, they have been kept alive in her memoirs, as well as many photo albums and scrapbooks, and now this book.

Mildred passed away on March 6, 1992, at the age of ninety-five. She shares a burial plot with her parents and grandparents in the Norton Cemetery in Kansas.

Mildred Boddy Reed (1896 - 1992)

The sign erected in honor of Colonel O.W. Reed at the crossroads in Villebaudon, France.
(Courtesy of Liliane Jamard, mayor of Villebaudon)

Acknowledgements

The research for this wide-ranging tour through the twentieth century required the assistance of many people and institutions. From individual family members to experts in their fields, this author deeply appreciates the contributions of all who are truly deserving of my gratitude for their contributions in bringing the stories of the Reeds and other families to light and life.

First and foremost are the descendants of Ollie and Mildred Reed: Bill Vaughan, Mark Reed, and Maryalyce Jenkins. Their support, assistance, and willingness to open their family history have been invaluable and beyond words of appreciation. Mildred Reed's niece, Patricia Tharp, added true gems of information and photos.

Veterans and families have been significant contributors to this story. Veteran of the 29th Division, Don McKee, conveyed the horrors he witnessed on July 30,1944. 91st Division veterans, Bob Palassou and Leon Weckstein, helped to bring to life the travels of their units from Oregon to North Africa and, finally, Italy. The support of a true Kansan and WWII vet, Senator Robert Dole, is deeply appreciated.

Sincere thanks go out to the families of those who served, many whom lost their lives in combat. Barbara Sall, daughter of Lester Maize, conducted thorough research into the events of July 30, 1944 in which Colonel Reed was killed, and Tom Bedecarre, whose wife's uncle, John Chittick, was killed the same day, were a tremendous help. Lillian Major, with assistance from Rev. Beth Palmer and David Rorick, and Rebecca Robbins for photographs of Clarence Major. Joe Bounds, with the help of Ed and Tom Northam, provided wonderful photos of his father, as did Barbara Fitzgerald of her father. It was a genuine honor to meet artist Luther Davis during his Honor Flight visit to the World War II Memorial in Washington, DC. Thanks to Mr. Davis's sister, Sarah Moss, for keeping us in touch, and a hearty thank you to the dedicated people with the Honor Flights. Jennifer Jackman, national chairman of the American Gold Star Mothers, heads an organization dedicated to supporting families suffering combat loss. Thanks to Abingdon Press for providing copies of *Strength for Service*.

Master Sergeant Roland Hall and Lt. Kevin Braafladt with the 91st Division have been supportive far beyond the call of duty, as have Sgt. Kelsey Sayson and Joe Balkoski, keepers of the 29th Division flame at the Maryland Military History Museum, and the 29th Division Association.

Thanks to Anne Gardocki, Jim Sloman, Pete Smith, Margaret Sloman, Steve Pryor, and Louise Delaney for Sloman family information. Laura Lepore and her daughters for help in the search for information about their mother and grandmother, Lenora Velie. Paula Frazier for sharing Nancy Campbell's photo album.

Special thanks to Brigadier General (USA, ret) Dr. Lance Betros, provost of the US Army War College, and former chair of the USMA Department of History for his review and comments on the West Point portions. Likewise, to the staff at West Point, Suzanne Christoff, Elaine McConnell, and Alicia Maulden. Appreciation to Anne Beiser Allen for photographs of her father at the academy. Thanks to the West Point Association of Graduates, and Suzanne Scullen-Ericson and her organization of 1942 graduates. A posthumous appreciation to General Willis G. Crittenberger.

Thanks to Greg Gardner, Susan Kilanski, and the staff of the Army Human Resources Command at Fort Knox for their perseverance.

At Wentworth, Al McCormick has been a huge help, along with Jim O'Malley, Susan Worthington of the Lexington Connection, Don Coen, and Jack Hackley in Lexington, MO, and posthumous thanks to Congressman Ike Skelton. In Norton, KS, thanks to Joe Ballinger of Norton Historical, Ardie Grimes, a Norton genealogist, and Mary Leuhers of the Norton Public Library.

In no special order but special thanks to Hal Buell; Courtney Bellizzi, Smithsonian Institution; Greg Palumbo; Shannon Brady; Luigi Benevnuti in Italy; Andre Shashaty; David Walsh; Daniel Brown, NPS Guam; Charlie and Emma Gallucio; Rev. Patrick Funston in Manhattan, KS; Father Michael Morris, Father Tom Devery, and the Archdiocese of New York; Anne-Marie Cutul in Scarsdale, NY; Jane Cruz in Delphi, IN; Patricia Patton and Cliff Hight at Kansas State University; Linda Glasgow and the Riley County Historical Society; Lisa Keys with the Kansas State Historical Society; Jill Hartke of the

State Historical Society of Missouri; Rev. Luther and Bobbie Davis-Miller; Jim Sheppard and the 50th Infantry Association; Marsha Smith of the Normandy Allies; Laura Tosi at the Bronx Historical Society; Patrick Raftery of the Westchester Historical Society; Sandra Reddish, historian, Fort Riley, KS; Shawn Faulkner, author, military historian, and educator; Dr. Peter Schifferle, author, military historian, and educator; Genoa Stanford at the Donovan Research Library, Ft. Benning; Zachary Hanner of the National Infantry Museum; Annie Wedekind for her help with the WWII Museum archives; James Boyle of the Pueblo County Library; the Pueblo County Historical Society; Jim Hurt of the National Railway Historical Society; S.H. Lustig and the New York Central Historical Society; Dennis Weaver in Merchantville, NJ; the staff of the Drexel University Library and Archives; Rebecca Jane Hamlett and Andrea Meloan at William Jewell College; Iowa Gold Star Museum; Des Moines Historical Society; Dick Wright in Leavenworth, KS and Elizabeth Dubuisson at CARL, Ft. Leavenworth; Herb Pankratz and Kathy Strauss at the Eisenhower Library; James Zobel with the MacArthur Memorial; Nathan Jones of the Patton Museum; author John McManus; Marcia McManus of the Army Chaplain Museum; Jim Currie; Luther Hanson, QMC Museum; French historian Antonin Dehays; Bob Hill; the Nash Car Club of America; and to my friends who have encouraged me along the way.

Special thanks to the hardworking and under-staffed members of the National Archives: Marie, Carpenti, William Cunliffe, Tim Nenninger, Mitch Yockelson, Patrick Osborn, Eric Vanslander, Theresa Fitzgerald, Wanda Devore, Sarah Forgey, Roy Lower, Joe Keefe, and, especially, the staff of the Still Pictures Branch.

Lindsey Alexander provided professional editing skills to the first draft and Andrea Wallis Aven held me to grammatical task. Author Al Barnes has been a stalwart colleague and friend throughout the research and writing. Jane Wright pitched in so many times to help with genealogy searches.

Last and certainly not least, my wife, Maryclare, for her patience and assistance throughout this long journey. Thank you!

Finally, an apology to anyone I have neglected to mention. Please know that your assistance and contributions are truly appreciated.

Arlington National Cemetery, 2011 (Dennis Whitehead)

Capt. Mac Patch, III

Omitted from the previous edition was a note about another young life lost in the war - Captain Alexander "Mac" Patch, III, son of General Alexander "Sandy" Patch. Mac graduated with Bud in the USMA class of 1942. Immediately after the graduation ceremony, he married Genevieve Spalding, daughter of Colonel Basil D. Spalding, in the Cadet Chapel. The newly-wedded Reeds and Patches reported to Ft. Benning two weeks later. From there, Mac was assigned to the 315th Infantry Regiment, 79th Division. He was killed in action on Oct 22, 1944 in Luneville, France and is buried in the Epinal American Cemetery.

Endnotes

CHAPTER ONE: NORTON, KANSAS

1 Anonymous, "Westward Ho!," Western Resources Monthly, April, 1901.

2 In 1980, Mildred Reed wrote and self-published, *Letters from Mit*, a memoir for the benefit of her grandchildren.

3 Norton *Courier*, November 16, 1899. Provided by Patricia Tharp.

4 The Homestead Act allowed settlers to claim 160 acres of public land for a small fee. If they lived on the land for five continuous years, built a residence, and grew crops, they could then acquire the deed to the property. They also could purchase the tract for $1.25 per acre after living on the land for six months, building a home, and starting to grow crops. By 1900, the act distributed 80 million acres of public land. Source: Kansapedia, Kansas Historical Society, http://www.kshs.org/portal_kansapedia

5 *Catalogue of the Officers and Students of the Upper Iowa University for the Academic Year, 1875-76*, Fayette County, Iowa, (Dubuque: Palmer, Winall & Co., 1876).

6 The last buffalo killed in Kansas was in April 1887, in Cheyenne County. Beccy Tanner. "Buffalo shaped the culture of Kansas," The Wichita *Eagle*, June 12, 2011. Read more at http://www.kansas.com/news/local/news-columns-blogs/the-story-of-kansas/article1066397.html#storylink=cpy

7 Dinah Faber. "The Dewey-Berry Feud." Wild West, 2003.

8 William Allen White. The Emporia *Gazette*. Full text available at the Kansas State Historical Society, http://www.kshs.org/kansapedia/what-s-the-matter-with-kansas/16717

9 Norton County *News*, August 24, 1916.

10 Mildred Reed, *Letters from Mit*, self-published, 1980.

CHAPTER TWO: WAR ON THE HORIZON

11 Text of unknown newspaper obituary provided by Patricia Tharp.

12 "History of the Kansas State University Army ROTC," http://armyrotc.k-state.edu/K-State-ROTC-History/index.htm

CHAPTER TWO: WAR ON THE HORIZON

13 Captain Lincoln C. Andrews, under the supervision of Major General Leonard Wood, *Fundamentals of Military Service*, J.B. Lippincott Company (not in copyright), 1916.

14 The "Zimmermann Telegram" was an encrypted diplomatic telegram sent

by German Foreign Minister Arthur Zimmerman to the German Ambassador in Mexico City on January 19, 1917. It was intercepted and decrypted by British naval intelligence. The "Zimmerman Telegram" promised the Mexican Government that Germany would help Mexico recover the territory it had ceded to the U.S. following the Mexican-American War, in return for an alliance against the U.S.

15   "The Mexican Border Problem," Kansas National Guard Museum, Topeka, KS.

16   C.H. Martin. "Rapid Transportation of Infantry," Infantry Journal, November-December 1916.

17   In more recent times, "selectees" would become known as "draftees."

18   Richard S. Faulkner, *The School of Hard Knocks: Combat Leadership in the American Expeditionary Forces* (College Station: Texas A&M University Press, 2012).

19   National Archives and Records Administration, Record Group 165: Records of the War Department General and Special Staffs. Administrative History, M1024, Roll 261, File 9226; Iowa City Press-Citizen, April 25, 1917.

20   Ibid., Faulkner.

21   Established in 1870 during the Indian Wars; the word "Benzine" was used in reference to the cleaning solution that, in this case, would wash away unsuitable officers.

22   Ibid., Faulkner.

23   Reverend Thomas J. Duvall's middle name was actually Jeffries.

24   National Archives and Records Administration, Record Group 407, Adjutant General Office, location 370:84/7/3-2, box 1083

25   Charles G. Hibbard, *Fort Douglas Utah 1862–1991*, (Fort Douglas: Fort Douglas Military Museum, 1999); Salt Lake City Telegram and Salt Lake City Tribune, November 10, 1917. Special thanks to Mr. Samuel Rogal.

26   The Salida *Mail*, Salida, Colorado, November 13, 1917.

27   Federal Bureau of Investigation, response to a Freedom of Information request, December 12, 2013. The Bureau of Investigation preceded the FBI.

28   File Number 485, Railroad: Denver & Rio Grande Railroad; Date: 11/12/1917; Location: Cotopaxi, CO. Interstate Commerce Commission, U.S. Department of Transportation, DOT Library, Investigations of Railroad Accidents 1911–1993.

29   The Ogden *Standard*, November 15, 1917.

30   Reed Smoot diaries, 1880-1932, Reed Smoot papers, 19th & 20th Century

Western and Mormon Americana, L. Tom Perry Special Collections, Harold B. Lee Library, Brigham Young University.

CHAPTER THREE: NEW LIFE BEGINS

31   Nancy W. Reifenstein, "The Influenza Epidemic at Camp Devens," Harvard, MA Town Clerk, July 1989. Fort Devens Museum.

32   Jim Duffy, "The Blue Death," Johns Hopkins Public Health, Fall 2004, http://www.jhsph.edu/news/magazine/archive/Mag_Fall04/prologues/index.html

33   The letter, written by Dr. Roy N. Grist to an unknown recipient named Burt, was discovered in a trunk and published in the December 22-29, 1979 edition of the British Medical Journal.

34   Carol R. Byerly, PhD, "The U.S. Military and the Influenza Pandemic of 1918-1919," Public Health Reports, 2010.

35   Alfred W. Crosby, *America's Forgotten Pandemic, the Influenza of 1918*, (New York: Cambridge University Press, 1989, 2003 edition); John M. Barry, The Great Influenza, (London: Penguin Books, London, 2004, 2009 edition).

CHAPTER FOUR: THE WAR TO END ALL WARS

36   Anne Leland and Mari-Jana Oboroceanu, "American War and Military Operations Casualties: Lists and Statistics," Congressional Research Service, 2010

37   "The Great Pandemic," United States Department of Health and Human Services, http://www.flu.gov/pandemic/history/1918/index.html

38   77th Division Association. *The History of the Seventy Seventh Division, August 25th, 1917 - November 11th, 1918*, (New York: W.H. Crawford Company, 1919).

39   Norval Dwyer, "The Camp Upton Story, 1917-1921," Long Island Forum, February 1970.

CHAPTER FIVE: GERMANY – THE ARMY OF OCCUPATION

40   Commonly spelled Koblenz today, Coblenz is a common English spelling and was used in pre-1926 Germany.

41   Alexander F. Barnes, "Coblenz 1919: The Army's First Sustainment Center of Excellence," Army Sustainment. September 1, 2010.

42   The *Associated Press* in The New York *Times*. "Pershing in Coblenz, Soldiers Line Streets," September 28, 1921.

43   Ibid., Barnes.

44   Edward M. Coffman, *The Regulars, the American Army 1898-1941*, (Cambridge: Harvard University Press, 2004).

45   Record Group 120, Records of the American Expeditionary Forces, Name Card Index to Special Orders, Y cables, Entry UD 36, Box 1, National Archives

and Records Administration.

46   Major Philip H. Bagby, Assistant Chief of Staff, G-2, *American Forces in Germany. American Representation in Occupied German, 1920-1921, Volume II.* U.S. Army. 1921.

47   Alexander Barnes, *In a Strange Land: The American Occupation of Germany, 1918--1923*, (Atglen, PA: Schiffer Military History, 2011).

48   Henry Hossfeld and Major Philip H. Bagby, Assistant Chief of Staff, G-2, *American Forces in Germany, American Representation in Occupied Germany, 1922-1923*, U.S. Army, 1923.

49   Ibid.

50   Ibid., RG 120, NARA.

51   Ibid., Hossfeld and Bagby.

52   Sturmabteilung, an early Nazi paramilitary group.

CHAPTER SIX: BETWEEN THE WARS

53   Interurbans were electric trolley lines connecting nearby communities. They were common in various parts of the country.

54   Col. S.D. Rockenbach, Inf., U.S.A. "Camp Meade: The Tank School at Camp Meade," Army and Navy Journal, April 1, 1922 (VMI Library).

55   Colonel and Major George S. Patton, "Tanks in Future Wars, Cavalry Journal, May 1920 and Infantry Journal, vol. 16, part 2,. 1920.

56   Captain Dwight D. Eisenhower, "A Tank Discussion," Infantry Journal, Vol. 27 (July-December 1920).

57   Conversation with Dr. Robert S. Cameron, historian, U.S. Army Armor Branch, Fort Benning, GA. April 16, 2014.

58   Richard W. Stewart, general editor, American Military History, Volume II: The United States Army in a Global Era, 1917-2003, (Washington, DC: Center of Military History, United States Army, 2005).

59   The reference to '17 harkens back to the "military cities" built in 1917.

60   First Lieutenant Leroy W. Yarborough, *A History of the Infantry School*, (Fort Benning: The Infantry School, 1931).

61   U.S. Senate Subcommittee on Military Affairs, Hearings on the Reorganization of the Army, December 4, 1919.

62   Army Appropriations Bill 1921, Subcommittee No. 1 of the Committee on Military Affairs, House of Representatives, March 25–April 2, 1920.

63   The *Doughboy* 1924, Infantry School yearbook, Fort Benning, GA.

64   Brigadier General Arthur L. Wagner (1853–1905) was an instructor at the

U.S. Infantry and Cavalry School, Fort Leavenworth, 1886–97.

65   The *Doughboy* 1923, Infantry School yearbook, Fort Benning, GA.

66   Sharyn Kane and Richard Keeton, *Fort Benning: The Land and the People*, (Tallahassee: Southeast Archaeological Center, National Park Service, 1998).

67   First Lieutenant Leroy W. Yarbrough and Major Truman Smith, *A History of the Infantry School, Fort Benning, Georgia, Volumes I and II*, (Fort Benning: U.S. Army, 1945).

68   John McAuley Palmer, *Washington, Lincoln, Wilson – Three War Statesmen*, (Garden City: Doubleday, Doran, 1930).

69   1924 *Doughboy*, Infantry School yearbook.

70   Ibid. A History of the Infantry School.

CHAPTER SEVEN: PROFESSOR OLLIE REED

71   Today, Drexel University in Philadelphia, PA.

72   Edward D. McDonald and Edward M. Hinton, *Drexel Institute of Technology, 1891-1941, A Memorial History* (Philadelphia: Drexel Institute of Technology, 1942).

73   Ibid., McDonald and Hinton.

74   Hollis Godfrey, *The Man Who Ended War* (Boston: Little, Brown, and Co., 1908).

75   "Dr. Hollis Godfrey, Engineer, 61, Dead," New York Times, obituary, January 19, 1936.

76   Mildred always referred to the house as being in Merchantville, NJ. Today, the house remains standing at 6159 Grant Avenue in Pennsauken Township.

77   The Drexel *Triangle*, Drexel Institute of Technology, May 28, 1928.

CHAPTER EIGHT: BACK TO BENNING

78   Ibid., *A History of the Infantry School*.

79   Though not directly related to the Confederate general of the same name, his father had served in the cavalry under Stonewall Jackson and named his son in the general's honor. Ollie Reed and Jackson served together in Germany with the 50th Regiment, as well as later at the Command and General Staff School at Fort Leavenworth. Both were promoted to captain on the same day, July 1, 1920. Major General Stonewall Jackson died in 1943 in a plane crash as commander of the 84th Infantry Division.

80   The Papers of George Catlett Marshall, ed. Larry I. Bland and Sharon Ritenour Stevens (Lexington, Va.: The George C. Marshall Foundation, 1981–). Electronic version based on The Papers of George Catlett Marshall, vol. 1. "The Soldierly Spirit," December 1880-June 1939 (Baltimore and London: The Johns

Hopkins University Press, 1981), pp. 319–321.

81   Ibid., Kane and Keeton.

82   Ibid., *A History of the Infantry School, vol. II.*

83   "Machine Gun Record Set by Company H 29th Infantry," Infantry School News, August 2, 1929.

84   Infantry School *News*, February 13 and 27, 1931, Fort Benning.

CHAPTER NINE: COMMAND AND GENRAL STAFF SCHOOL

85   American Military History, Center for Military History, U.S. Army, Washington, DC, 1989.

86   Ibid., Coffman.

87   Today, Command and General Staff College at Fort Leavenworth.

88   National Archives and Records Administration, Record Group 407, Location: 370:81/13-14/7-2, Box 1914.

89   Edward M. Coffman and Peter F. Herrly, "The American Regular Army Officer Corps between the World Wars," Armed Forces and Society, November 1977.

90   Ibid., Coffman.

91   *Command and General Staff College, Fort Leavenworth, Kansas*, Command and General Staff College (CGSC), U.S. Army, Fort Leavenworth, KS, 1950.

92   Ibid.

93   Col. Elvid Hunt, *History of Fort Leavenworth, 1827–1937, Command and General Staff School, 1937*; 1981 reprint, Fort Leavenworth Historical Society.

94   Dr. Michael J. King, *Brief history of the US Army Command and General Staff College, 1881-1981*, Combat Studies Institute, USAC & GSC, Fort Leavenworth, KS, 1981, 19-23.

95   *Military history of the U.S. Army Command and General Staff College*, Fort Leavenworth, Kansas 1881-1963, Command and General Staff College, Fort Leavenworth, KS, 1963.

96   Ernest Nason Harmon and Milton MacKaye, *Combat Commander: Autobiography of a Soldier*, (Englewood Cliffs, NJ: Prentice-Hall, 1970).

97   National Archives and Records Administration, Record Group 407, Adjutant General Office, Location 370:81/13-14/7-2, Box 1914.

98   Forrest C. Pogue interview with Rev. Luther D. Miller at the National Cathedral, Washington, DC, June 21, 1962, cited with permission of the George C. Marshall Foundation.

99   Correspondence with Alexander Barnes, October 30, 2012.

Love and Sacrifice

100 Schifferle, Peter J., *America's School for War: Fort Leavenworth, Officer Education and Victory in World War II*, (Lawrence: University Press of Kansas, 2012).

101 GS – General Staff, G-1 Personnel, G-2 Intelligence, G-3 Plans and Training, and G-4 Logistics.

102 October 2012 correspondence with Dr. Peter J. Schifferle, Professor of History, Command and General Staff College, Fort Leavenworth.

CHAPTER TEN: BACK TO SCHOOL, FOR BUD'S SAKE

103 Correspondence, December 31, 2012, Mike Floberg, P.E., Chief, Bureau of Transportation Safety and Technology, Kansas Department of Transportation, http://www.ksdot.org/burtransplan/maps/HistoricStateMaps.asp

104 Raymond W. Settle, *The Story of Wentworth: History of Wentworth Military Academy, Lexington, Missouri 1880-1950* (Lexington: Wentworth Military Academy, 1950).

105 Ibid., Settle.

106 Riley's collapsed into a pile of rubble during the summer of 2013, just hours before the owner, Katherine Van Amburg, lost her battle with cancer.

107 Correspondence, James O'Malley to Jack Hackley, February 2, 2013.

108 Unusually high waters in November 1935 finally washed away Bootleggers Island.

109 One bright side to the grasshopper infestation was that local turkeys grew fat eating them.

110 National Archives and Records Administration, Record Group 394, Records of United States Army Continental Commands, 1920-1942, Entry 205, Location 82:14/04, Box 3.

111 The *Red Dragon*, Wentworth Military Academy, December 2012.

112 National Archives: Record Group 394, Entry 205, Location 82:14/04 Box 3, Folder 314.7.

113 Subsequent years would be judged by fellow Hollywood stars Robert Taylor, Fred McMurray and Tyrone Power.

114 Correspondence with The Lexington *Connection*, January 2014.

115 Greene, Lorenzo, Kremer, Gary and Holland, Antonio, *Missouri's Black Heritage*, Columbia: University of Missouri Press, 1993.

116 Missouri ex Rel. Gaines v. Canada, Registrar of the University of Missouri, et. al. Supreme Court of the United States, 305 U.S. 337, November 9, 1938, Argued, December 12, 1938, Decided. See http://gobcc.missouri.edu/about/history/ for information about the Gaines/Oldham Black Culture Center at the

University of Missouri.

117 Leslie H. Bell, *Educational Heritage of a Century: History of the Lexington Public Schools*, Board of Education, R-V School District, 1962.

118 Andrew Dubill, council historian, Boy Scouts of America, Heart of America Council, Kansas City, Mo.

CHAPTER ELEVEN: DUTY, HONOR, COUNTRY: WEST POINT

119 Today this is the Eisenhower Executive Office Building, an adjunct of the White House where the offices of the vice president and other executive branch offices are located.

120 U.S.C.C. Memorandum No. 76, June 8, 1937, Section VIII, Acceptance and Enrollment of New Cadets, Par. 32-35, United States Military Academy at West Point.

West Point classes are defined numerically the opposite of civilian college, each with a specific name. There are no freshman, sophomore, junior, and senior classes; rather, first-year cadets are Plebes, also called Fourth Classmen; the next class up are Third Classmen or Yearlings; then the Cows or Second Classmen; and finally are the Firsties – the First Classmen, the graduates-to-be.

121 Bill Yenne, *Black '41 – The West Point Class of 1941 and the American Triumph in WWII*, (New York: John Wiley & Sons, 1991).

122 The Army tradition of no letter J dates to eighteenth- and nineteenth-century script. J was too similar to I, so it was omitted from official designation.

123 USMA class of 1918, Major General Edwin L. Sibert, assistant Chief of Staff, European Theater of Operations.

124 Theodore J. Crackel, *The Illustrated History of West Point*, (New York: Harry N. Abrams, 1991).

125 Issue demerits

126 Walking tour punishment.

127 William R. Buster, *Time on Target: The World War II Memoir of William R. Buster*, (Frankfort: Kentucky Historical Society, 2001).

128 Ibid., Yenne.

129 "United States Military Academy at West Point," Thomas Jefferson, Foundation, Inc. Charlottesville, VA.

130 Stephen Ambrose, *Duty, Honor Country – a History of West Point*, (Baltimore: Johns Hopkins University Press, 1999).

131 Sidney Forman, *West Point*, (New York: Columbia University Press, 1950).

132 Ibid., Crackel

133 Ibid., Crackle.

134 John Grant, James Lynch, and Ronald Bailey, *West Point – The First 200 Years*, (Guilford: Globe Pequot Press, 2002).

135 Ibid.

136 Ibid, Ambrose.

137 Cadet F. W. Ebey, "Some 'Kaydets' Enjoy Dress Parade; Average Man Doesn't, Writes Pointer," Harvard Crimson, October 18, 1930.

138 Brewerton, Hank (Class of 1940). "Assembly, Association of Graduates, U.S. Military Academy," December 1976, page 126, West Point Library.

139 John Eisenhower, *General Ike: A Personal Reminiscence*, (New York: Simon and Schuster, 2004).

140 Recollections of General Willis Crittenberger, Jr. in June 2013 conversation.

141 Franklin D. Roosevelt: "Graduation Address at United States Military Academy, West Point, New York," June 12, 1939, The American Presidency Project, John T. Woolley and Gerhard Peters, University of California, Santa Barbara.

142 "West Point Parades Abandoned," The New York *Times* (AP), April 6, 1940.

CHAPTER TWELVE WELCOME TO THE WORLD OF TOMORROW

143 "Your World of Tomorrow," 1939 World's Fair booklet, Gilbert Seldes, Rogers-Kellogg-Stillson, Inc. 1939.

144 NYPL Image 1679595

145 Dwight Macdonald, "Reading from Left To Right," New International, Vol.5 No.7, July 1939, pp.221-223.

146 See the Museum of the City of New York website, http://mcnyblog.org/tag/amusement-parks/, for an interesting overview of the fair, as well as http://www.1939nyworldsfair.com/.

CHAPTER THIRTEEN: SAILING TO A NEW HORIZON

147 The Brooklyn Army Terminal was a five-million-square-foot military embarkation terminal built in just 17 months during World War I. The facility gained particular notoriety in September 1958 when Private Elvis Presley departed from the Brooklyn terminal for his Army assignment in Germany.

148 Inaugurate in 1935, the China Clipper was a four-engine flying boat built for Pan American Airways for commercial travel from San Francisco to Manila. Between the two destinations, the Clipper has stops in Honolulu, Midway Island, Wake Island, and Guam. Trans-Pacific mail was, perhaps, the China Clipper's most important cargo.

CHAPTER FOURTEEN: THE PHILIPPINE ISLANDS

149 MacArthur is best remembered during this period for his heavy-handed treatment of World War I veterans known as the Bonus Army, who marched on Washington, DC, seeking bonus payments owed them. MacArthur employed infantry, cavalry, and tanks to drive the veterans and their families out, burning their encampments to the ground.

150 Hiroshi Masuda, *MacArthur in Asia: The Generals and His Staff in the Philippines, Japan, and Korea*, (Ithaca, NY: Cornell University Press, 2012).

151 Kris or keris – a wavy-edged dagger popular in the southern Philippines.

152 A Moro swordsman on a personal mission of jihad.

153 Courtesy of the Smithsonian Institution Archives, Theodore Reed Oral History Collection, RU 9568, Interview 1.

CHAPTER FIFTEEN: WAR CLOSES IN

154 Louis Morton, *United States Army in World War II, The War in the Pacific, The Fall of the Philippines* (Washington, DC: U.S. Army, Center of Military History, 1953).

155 Brigadier General Albert Monmouth Jones was taken prisoner and later awarded the Distinguished Service Cross, the Distinguished Service Medal, Silver Star, Legion of Merit and POW Medal. Lieutenant Colonels Doane, Fortier, Crews; Majors Townsend Vesey and Uhrig were all taken prisoner by the Japanese in the defense of the Philippines and survived the Bataan Death March. Major Vesey, a 1918 West Point graduate, died in captivity. Lieutenant William C. Fite, a 1938 graduate of West Point, went on to win the Silver Star.

156 Courtesy of the Smithsonian Institution Archives, Theodore Reed Oral History Collection, RU 9568, Interview 1.

157 Ibid. Lowe.

CHAPTER SIXTEEN: LAURA

158 *Presidential Campaign Expenses, Hearing Before a Subcommittee of the Committee on Privileges and Elections*, U.S. Senate, Sixty-Sixth Congress, convened September 2, 1920.

159 Neither of these could be confirmed.

160 *Life* magazine, April 14, 1941, pages 32-33, photographs by Peter Stackpole and Herbert Gehr.

161 Robert Keith, "Initial Federal Budget Response to the 1941 Attack on Pearl Harbor," CRS Report for Congress, (Washington, DC: Congressional Research Service, 2001).

162 Dr. Stetson Conn, *Highlights of Mobilization, World War II, 1938-1942*,

Love and Sacrifice

Office of the Chief of Military History, (Washington, DC: U.S. Army, 1959).

163 The Selective Training and Service Act of 1940, also known as the Burke-Wadsworth Act.

164 "Speech to the Graduating Class, United States Military Academy," 1942, Document #3-205, George C. Marshall Foundation, Lexington, VA

CHAPTER SEVENTEEN: INTO THE FRAY

165 Roy Livengood, *Powder River! A History of the 91st Infantry Division in WWII*, Paducah, KY: Turner Publishing Company, 1994).

CHAPTER EIGHTEEN: DEPARTURES

166 Recollections of Roy Brown who served aboard the Liberty ship SS John Dickinson in the convoy. Mackenzie J. Gregory, http://ahoy.tk-jk.net/macslog/ConvoyUGS-40underextremee.html.

167 Leon Weckstein, *Through My Eyes, 91st Infantry Division in the Italian Campaign, 1942-1945*, (Ashland, OR: Hellgate Press, 1999), and correspondence with Mr. Weckstein.

168 In 2012, Ollie Reed, Junior's son, Bill Vaughan, had the opportunity to meet and thank Mr. Davis during an Honor Flight of South Carolina veterans to the World War II Memorial in Washington, DC.

169 West Point Association of Graduates, Memorials, John H. Van Vliet 1913.

CHAPTER NINETEEN: SAFE LANDINGS

170 Adrian and Neil Turley, *The U.S. Army at Camp Bewdley and Locations in the Wyre Forest Area, 1943 – 1945*, (Bewdley, UK, A&N Turley, 2000).

171 General George S. Patton Museum and Center of Leadership, Fort Knox, KY; Patton delivered this speech to several different army groups in preparation for the invasion of Europe.

172 Colonel William H. Cornog, Jr. was the commanding officer of the 36th AIR – Armored Infantry Regiment, 3rd Armored Division. He was a football legend and 1924 graduate of West Point. Cornog was killed in an August 10, 1944 artillery attack on his command post at Juvigny, France.

173 It was June 20.

174 Interview of Gen. Charles Bolte (OH-395) by Dr. Maclyn Burg of the Eisenhower Library on January 29, 1975.

175 *Strength for Service to God and Country*, edited by Chaplain Norman E. Nygaard, copyright Whitmore & Stone, 1942, Abingdon-Cokesbury Press. New York and Nashville.

176 Colonel Paul Ryan "Pop" Goode was a leader among prisoners held at the POW camp in Moosburg, Germany, for which he received a Silver Star. He

retired from the Army in 1952, and was named deputy Governor of the U.S. Soldiers Home in Washington, DC.

177 Military Intelligence Service, War Department, EX-Report No. 617, May 17, 1945.

178 29th Infantry Morning Report Index, transcribed by Ralph Windler, Maryland Military Historical Society, Baltimore.

179 Informal conversation between Captain John Kernan Slingluff and his father, Jesse Slingluff, Sr., Baltimore, Maryland, 1945, kindly provided by Shirley Isbill.

180 29th Infantry Division Order of Battle; Joseph Balkoski, *Beyond the Beachhead: The 29th Infantry Division in Normandy*, (Mechanicsburg, PA: Stackpole, 1989). Joseph Ewing, *29th Infantry Division: A Short History of a Fighting Division*, (Nashville: Turner, 1992).

181 Ibid, Balkoski.

182 Ibid., Balkoski.

183 Courtesy of the Smithsonian Institution Archives, Theodore Reed Oral History Collection, RU 9568, Interview 1.

CHAPTER TWENTY: ITALY

184 Chester B. Hansen collection, 1928-1952, undated, U.S. Army Heritage & Education Center, Carlisle, Pennsylvania

185 Sidney T. Mathews, "General Clark's Decision to Drive on Rome," chapter fourteen of Command Decisions, edited by Kent Roberts Greenfield, (Washington, DC: Center for Military History, 1960).

186 Correspondence with Bob Palassou, veteran of the 363rd Regiment, July-August 2011.

187 *91st Division, August 1917 – January 1945*, Information Education Section, MTOUSA, 1945, Lone Sentry, www.lonesentry.com; Ibid., Weckstein.

188 Correspondence with Bob Palassou, veteran of the 363rd Regiment, 2013.

189 Individual Deceased Personnel File (IDPF), U.S. Army Human Resources Command, Fort Knox, KY, Susan M. Kilianski and Melissa McCartney.

190 Captain Ralph E. Strootman, *History of the 363rd Infantry*, (Washington, DC: Infantry Journal, 1947).

191 Record Group 407, Records of the Adjutant General's Office, 391-INF(363), National Archives and Records Administration.

192 The date of Bud's death is incorrectly etched upon his cross in the Normandy American Cemetery as July 6, 1944. His place of birth is recorded as Missouri, not Connecticut.

Love and Sacrifice

193 Ibid., Palassou correspondence.

194 MSG Roland Hall, 91st Division, Historic Collection Manager, Ft. Hunter Liggett, CA.

CHAPTER TWENTY-ONE: NORMANDY

195 Ibid., Balkoski.

196 The Maryland National Guard Military Historical Society, Fifth Regiment Armory, Baltimore, MD, 29th Division After-Action Reports, War Room Journal; ibid, Balkoski.

197 *American Forces in Action Series: St-Lô (7 July-19 July 1944)*, Historical Division, War Department, 1946, Washington, DC.

198 George Wilson, *If You Survive: From Normandy to the Battle of the Bulge to the End of World War II, One American Officer's Riveting True Story*, (New York: Ballantine Books, 1987).

199 Scarsdale *Inquirer*, "Army Chaplain is Decorated, Saw Ten Wounded Murdered by Foe," October 13, 1944.

200 Barbara Sall, "Action of the 175th", personal research document.

201 National Archives and Records Administration, National Personnel Records Center, 201 file, Taggart, Gerard W., 488189.

202 Letter from Don McKee to Regimental Surgeon Dr. Edmund G. Beacham, Dr. Levin and Dr. Chambers: "Tragic Action occurring in the 175th area 30 July 1944 – A sad day for the Geneva Convention." Thanks to Barbara Sall for providing a copy.

203 Extracts from *Saint-Lô (7 July-19 July 1944), American Forces in Action Series*, Historical Division, War Department, 21 August 1946; Ibid. Balkoski.

CHAPTER TWENTY-TWO: LOVE AND LOSS

204 John C. McManus, *The Dead and Those About to Die: D-Day: The Big Red One at Omaha Beach*, (New York: NAL Caliber, 2014).

205 Today, the location is the Gramercy Mansion, a Baltimore County bed and breakfast, that still operates the Koinonia organic farm that was part of the original center.

206 From the notes of Mildred's cousin, Patricia Tharp.

207. David Grayson, *The Friendly Road: New Adventures in Contentment*, Grosset & Dunlap, 1913

PHOTOGRAPHS

Photographs appearing in *Love and Sacrifice* primarily originated in the Reed and Sloman family collections, as well as Patricia Tharp. Other sources are noted.

Captain O.W. Reed, seated front and center, Quartermaster Corps, Germany - undated.

Index

A

Abadan Air Base 202
abolition 8, 113
Adams, President John 148
Alexander, General Sir Harold R. L. G. 253
Alexander, Sergeant Guy B. 35
Algeria Arzew/Oran/Port aux Poules 231, 232, 237
Allegheny Cemetery 199, 291
Allen, Major General Henry T. 54, 63
Almquist, Captain Elmer H. 106
Almquist, Mary 105
American Expeditionary Forces (AEF) 52, 71
American Forces in Germany (AFG) 52, 53, 54, 56, 57, 62, 65, 108
Anderson, Robert J. 281
Angel, Private Clyde T. 281
Antwerp, Belgium 53, 55, 66
Anzio, Italy 258, 259, 287
Arlington National Cemetery 301
Armistice 51, 52, 101, 115
Army Air Corps 95, 202, 301
Army-Navy game 122, 152, 204
Army pin ('A' pin) 204
Army War College 102, 103, 109, 306
Arnold, General Henry "Hap" 159
Arno River 260
artillery 27, 37, 73, 105, 151, 247, 248, 252, 259, 260, 262, 263, 265, 267, 268, 269, 270, 272, 277, 279, 280, 286, 319; coastal artillery 27, 73, 105
Aull 112, 114, 137, 300
Aull, Sarah Ann 300
Auslander, Joseph 273
Austro-Hungarian Empire 51, 64
Ayer, Massachusetts 39, 42

B

B-17 276
B-24 276, 301
Bagnoli, Italy 243, 253, 258, 259
Baker, Private William Noah 264
balut 187
Barbour, Ann 202
Barnes, Alexander 108, 312, 315
Bavaria 51, 64
Bavarian Soviet Republic (Bayerische Räterepublik) 51
Bayeux-Saint-Lô highway 272
Beast Barracks / Cadet Basic Training 141, 142, 146, 147, 151, 161
Beauregard, Pierre G.T. 151
Beer Barrel Polka 218, 275
Beiser, Jay 143, 158, 306
Belgium 51, 52, 53, 56, 66
Bend, Oregon 221
Bendorf, Germany 60, 61
Benedict, Brigadier General Jay 162
Benzine Board 27
Berlin, Germany 64, 273
Berlin, Irving 50
Berry, Margaret 107
Beverly, New Jersey (Beverly Farm) 208, 215
Bingham, Major Sidney 271
Black Tuesday 99
Blackstone, Virginia 222
Blitzkrieg 168, 205
Bluemont Avenue, Manhattan, Kansas 283
bocage 245
Boddy, Bert 6
Boddy, Della Sarvis 6, 10, 11, 12, 16, 18
Boddy, Mildred (*See also* Reed, Mildred Boddy) 3, 5, 10, 14, 28, 303

Boddy's Baptist Builders 12
bolo 177
Bolte, General Charles L. 245
Bone, Second Lieutenant Joseph W. 256
Borgstrom brothers 302
Boston 39, 40, 152, 313
Bounds, Sergeant Vaughn 279
Bowen, Lieutenant Colonel Millard G. 278
Bradley, Mary 94
Bradley, General Omar N. 92, 94, 108, 120, 252
Branch Night 211
Brewer, Pamela 294
Bright's Disease 199
Brisbane, Arthur 202
Bronx, New York 160, 166, 201, 202, 307
Bronze Star 266
Brooklyn Army Terminal 169, 317
Bryan, William Jennings 21
Bugle Notes 144
Bulgaria 64
Burlington, New Jersey High School 204
Burnham, Daniel H. 174
Bushwhackers 8

C

Cabanatuan 300
Cadet Basic Training (See Beast Barracks)
Cadets (Kaydets) 156, 317
Cain, Private First Class Gathier Howard 264
Calculator (Calc/Cal) 79, 80
calesa 184, 185
Camm, General Frank 288
Camp Adair 221, 223
Camp Bewdley (Bewdley, UK) 241, 319
Camp Butner 212, 221, 225, 287
Camp Devens 39, 40, 41, 42, 47, 48, 49, 311
Camp Dix 50
Camp Dodge 32, 35, 36, 37, 38, 48
Camp Meade 70, 71, 72, 87, 241, 312
Camp Patrick Henry 225, 226
Camp Pickett 222, 225, 235, 287
Camp Upton 47, 48, 49, 50, 311
Camp White 218, 219, 220, 221
Campbell, Lola Frazier 130
Campbell, Nancy Elizabeth 130, 131, 132, 133, 135, 137, 156, 161, 162, 171, 195, 196, 203, 211, 234, 257, 258, 296, 306, 311
Campbell, Ralph W. 130
Canham, Colonel Charles D.W. 268
Cantonment 77
Captain Cook 179
carabaos 177, 178
Caroline Islands 192
Casa de Alfabetizacion 292, 293, 294
Cebu 180
Cecina, Italy / Cecina River 259, 260
cemetery 129, 245, 274, 290
Chattahoochee Choo-Choo 75
Chattahoochee River 74
Chesapeake Bay 73, 160, 226
Chianni, Italy 260, 261, 262, 263
Chicago, Illinois 6, 26, 76, 107, 118, 119, 164, 165, 174, 200, 201, 202, 264
Chicago Herald & Examiner 200, 202
Chick, Captain Opie E. 279
China (Chinese Nationalists) 190
China Clipper 317
Chittick, Lieutenant John W. 279
Churchill, Prime Minister Winston 205
Church of Jesus Christ of Latter-day Saints (LDS) 35
Citizens Military Training Corps (CMTC) 202
City Beautiful Movement 76, 174

City College of New York 202
Civil War 6, 8, 21, 27, 34, 71, 78, 103, 113, 127, 134, 150, 151
Civitavecchia, Italy 258, 259
Clark, General Mark 253
Clay, General Lucius 159
clearing station 280
Cleveland, Ohio 198, 201
Coast Guard 160, 229, 230
Coblenz, Germany 52, 53, 54, 55, 56, 57, 65, 66, 311
Coffin, Howard 85
Collins, Brigadier General Edgar T. 92
Collins, General J. Lawton "Lightning Joe" 92
Colorado 5, 26, 30, 33, 34, 35, 158, 310
Columbia University Teachers College 198
Columbus, Georgia 81, 89, 218
Columbus, New Mexico 24
Combat Infantryman's Badge 285
Command and General Staff School (CGSS) 98, 99, 100, 219, 244, 289, 313, 314
Committee on Public Information 197
Committee to Defend America 190
Communist 51
Communists 64, 191
Conley, Edgar T. 139
Conley, Major General Edgar T. 139
Connecticut 39, 48, 49, 50, 128, 164, 202, 290, 320
Connecticut Yankees 202
Connelly, Congressman John Robert 19
Connor-Wagoner store 129
Conservation Biology Institute 297
convoy 229, 231, 319
Cook, Major General Gilbert R. 242
Cooper, Gary 123
Cooper, Lieutenant Colonel John P. 251
Corlett, General Charles H. "Cowboy Pete" 248
Cormoran 31
Cornog, Colonel William "Jug" 243
Corps of Engineers 149, 211
Corregidor 193
Corvallis 221
Cotopaxi 33, 310
Council of National Defense 85
Country Captain 186
Cows 178, 316
Cox, James M. 197
Craig, General Malin 100, 139
Creel, George 197
Crews, Lieutenant Colonel Leonard R. 192
Cristobal 170
Crittenberger, Willis 159
Crocker, David 160
crossroads 245, 248, 268, 277, 279
Crowden, Captain Eugene E. 232, 238, 262, 263, 264
Cuartel de España, Fort Santiago, Philippines 173
Cumming, Captain Samuel C. 106
Cunningham, Aulene C. 215
Custer, General George 107

D

Davis, Jefferson 150
Davis, Luther 233, 299, 305
Dayton, Ohio 197, 198
D-Day 238, 239, 244, 245, 247, 250, 274, 321
Deeley, Maud 29
demobilization 50, 51, 73
Democracity 165, 167
Dentler, Colonel Clarence E. 35
Denver & Rio Grande Railway 34, 35
Des Moines, Iowa 37, 307
Dewitt Clinton High School, Dallas, Texas 202
Dicken, James Fletcher 296
Dinky Line 75, 77

Disney, Walt 166
Distinguished Flying Cross 301
Doane, Lieutenant Colonel Irvin E. 192
Doughboy 75, 76, 78, 80, 82, 312, 313
Dreamland 116
Drewry, Jim 98
Drexel Institute of Technology (Drexel University) 82, 83, 313
Drought 119
Duchess Nebbie 220, 221, 240, 256
DUKW (ducks) 231
Dust Bowl 111
dust storms 119, 129
Duvall, Della 18 (*See also* Boddy, Della Sarvis)
Duvall, Reverend Thomas Jeffries 11, 28, 293

E

Eagle Pass, Texas 24
Eastern Shore, Maryland and Delaware 241, 302
East Orange, New Jersey 196, 202
Ederle, Gertrude 203
Edgemont, Westchester, New York 201
Edge, Senator Walter 198
Ehrenbreitstein Castle, Coblentz, Germany 66
Eisenhower, Dwight D. 71, 72, 81, 108, 120, 159, 175, 176, 252, 307, 312, 316, 317, 319
Eisenhower, Milton 234
England 6, 134, 190, 199, 213, 231, 232, 234, 239, 240, 241, 245
English Channel 203, 239, 243, 290
Ermita District, Manila 174

F

Farley, Henry S. 151
Fayro Laboratories 199
femmes 141, 151, 152, 196
Field Artillery 31, 144, 251, 280

Filipino 173, 175, 184, 185, 186, 206
Filipino Constabulary Band 173
Filipinos 174, 176, 191
Firstie 210
First Officers' Training Camp (FOTC) 25, 26, 30
First World War / Great War / World War I 51, 84, 99, 101, 103, 170, 219
Fitzgerald, Lieutenant Curtis 279
Flanders, Belgium 272
Flipper, Henry O. 151
Flirtation Walk 162, 196, 197
Florence, Italy 58, 243
flying lessons 256, 291
football 13, 14, 22, 54, 80, 81, 86, 87, 93, 98, 115, 116, 122, 123, 137, 152, 156, 162, 204, 209, 243, 319
Ford, George B. 93
Formosa 192
Fort Benning 74, 75, 76, 77, 79, 80, 81, 82, 86, 89, 91, 92, 93, 94, 95, 96, 97, 99, 126, 141, 163, 164, 174, 216, 218, 312, 313, 314
Fort Bliss 220
Fort Douglas 28, 30, 31, 48, 310
Fort Howard 73
Fort Jackson 206, 207
Fort Leavenworth 98, 99, 100, 102, 104, 105, 107, 110, 127, 135, 139, 290, 313, 314, 315; Post Chapel 93, 106, 107, 290
Fort Riley 19, 26, 136, 291, 307
Fortier, Lieutenant Colonel Malcolm 192
Forward Looking Association (Let's Redeem Ohio) 197
FOTC (*See* First Officers' Training Camp)
Fourth of July 252
France 39, 42, 47, 50, 51, 52, 53, 56, 64, 67, 71, 72, 78, 100, 108, 134, 161, 168, 190, 213, 231, 232,

240, 241, 244, 245, 265, 266, 279, 281, 284, 285, 288, 290, 319
Frazier, Lola 130
Free, Machinist's Mate Thomas Augusta 301
Free, Seaman William Thomas 301
Frisbie, John J. 9
Front Royal, Virginia 297
Funston, Major General Frederick 24
Fuqua, Major General Stephen O. 100
furlo 178

G

Garramore, Sergeant Vincent 234
Gaspard, Richard 160
Gasser, Colonel Lorenzo D. 100
General Electric 163, 167
General Staff 98, 99, 100, 109, 206, 219, 244, 289, 313, 314, 315
General Staff Eligible List 109
George, Lieutenant Colonel Alexander 247
Georgia 74, 75, 77, 81, 89, 94, 218, 313
Gerhardt, General Charles H. 108, 109, 124, 219, 244, 245, 246, 248, 249, 252, 266, 267, 268, 269, 270, 271, 273, 274, 275, 276, 277, 279, 281, 285, 286; Uncle Charlie 245
German 88-millimeter gun (German "88" / Blues 88s) 88, 250
German Aircraft
  FW Kodor 231
German Army /Wehrmacht
  2nd Panzer Division 277, 278
  XIV Panzer Corps 261
  19th Luftwaffe Field Division (Luftwaffen-Feld-Divisionen, LFD) 261, 262
  60th Panzer Grenadier Division 277, 278
  116th Panzer Division 277
Germans 31, 42, 50, 55, 56, 60, 62, 63, 65, 66, 67, 231, 232, 246, 252, 253, 260, 262, 263, 265, 270, 271, 272, 277, 278
Germany 21, 22, 31, 50, 51, 52, 53, 54, 55, 57, 59, 60, 63, 64, 65, 67, 68, 70, 74, 96, 108, 160, 166, 168, 190, 205, 207, 240, 281, 291, 310, 311, 312, 313, 317, 319
Gillis, Ann 201, 202
Gillis, Major William G. 212
Gill, Lieutenant Colonel Edward 279
Godfrey, Dr. Hollis 84, 313
Goode, Colonel Paul R. "Pop" 246, 247, 319
Gordon, William 160
Göring, Reichsmarshall Hermann 261
Gothic Line 260, 264
Grant, General Ulysses S. 103, 150
Graves Registration (Graves men) 55, 264
Great Britain 52, 168, 205
Great Depression 99, 101, 118, 124, 165, 197, 201; Depression-era 167
Great War (*See also* First World War/World War I) 22, 47, 69, 108, 170
Griffiths, Ann Quarles 300
Griffiths, Kenneth 158, 300
Grist, Dr. Roy N. 41, 311
Guam 31, 171, 172, 190, 306, 317

H

Hampton Roads Port of Embarkation 226
Hanner, Zachary Frank 93
Hansen, Major Chester 252
Harding, Edwin F. 92
Harding, Major Edwin F. 92
Harding, President Warren G. 64, 65, 198
Harmon, General Ernest Nason 104
Harness, Major Leslie 248

Hays, Captain James 277
Hearst, William Randolph /
    Hearst-Brisbane Properties /
    Hearst Corporation 202
hedgerows 245, 249, 267, 269, 270, 276
Hemingway, Ernest 200
Henson, Pamela 192
Hewitt, Jay 215
Highland Falls, New York 135, 139, 140, 162
Hill 108 (Purple Heart Hill) 247, 248
Hitler, Adolph 64, 68, 137, 166, 168, 276
Hodges, Major General Courtney H. 233
Hollister, Nellie 39, 48, 49, 164
Honolulu / Pearl Harbor 170, 193, 317
Hosea, Roland 294
Howie, Major Tom 271
Huebner, Major General Clarence 289
hump 100, 101

I

Immaculata High School, Leavenworth, Kansas 106, 119
Infantry Journal 71, 72, 310, 312, 320
Infantry School 74, 75, 76, 77, 78, 79, 80, 81, 82, 91, 92, 93, 94, 95, 96, 98, 102, 105, 106, 141, 216, 218, 312, 313, 314
Infantry School News 79, 80, 94, 95, 314
influenza 40, 41, 42, 47, 49, 67
Intramuros, Manila, Philippines 173
Iowa 6, 32, 37, 48, 75, 198, 307, 309, 310
Ireland, Surgeon General Merritte W. 40
isolationism 65, 124

J

Jackson, Captain Stonewall 92, 106, 313
Jackson, Governor Claiborne Fox 113
Jackson, General Thomas J. "Stonewall" 150
Japan / Japanese 34, 134, 137, 166, 175, 176, 189, 190, 191, 192, 206, 207, 223, 291, 300, 301, 318
Jayhawkers (Jayhawks) 8, 59
Jefferson, President Thomas 29, 49, 148, 149, 150, 293, 316; Mr. Jefferson's Army. 149
Jenkins, Maryalyce Reed 295, 298, 305
Johnsen, Private First Class Russell J. 264
Jolo 180
Jones, Colonel Albert M. 192
June Week 211

K

Kaltenborn. H.V. 165
Kansas 3, 5, 7, 8, 9, 10, 11, 18, 19, 20, 21, 22, 24, 25, 26, 55, 56, 58, 59, 62, 73, 82, 98, 99, 101, 102, 104, 111, 112, 113, 117, 118, 119, 121, 123, 124, 129, 133, 135, 156, 158, 185, 193, 206, 207, 219, 235, 240, 251, 255, 283, 291, 293, 296, 297, 298, 300, 302, 306, 309, 310, 314, 315, 316
Kansas City 7, 20, 98, 112, 117, 118, 121, 123, 124, 129, 133, 135, 156, 296, 298, 316
Kansas City Zoo 298
Kansas National Guard 24, 25, 26, 310
Kansas-Nebraska Act 113
Kansas State University (KSU/K-State) 19, 20, 21, 22, 56, 104, 193, 240, 251, 297, 298, 306, 309
Kansas State Agricultural College (KSAC) 20, 21, 22, 56
Karlstad, Captain Charles H. 105
Kashdan, Isaac 155
Kemper Military School 122

Kenyon, Senator William S. 198
Kilbourne, Brigadier General Charles E. 100
King, Brigadier General Campbell 93
King George VI and Queen Elizabeth 161
Koinonia 293, 321
König Wilhelm II 170
KSAC 20, 21, 22
Kuhlmann-Anderson Tests 204
Kuna, Idaho 292, 294, 302
Kyser, Kay 212

L

Laacher See, Germany 52
La Cambe, France 274, 281
La Denisiere, France 277
La Madeleine, France 271
Lamarr, Hedy 203
Laubach, Frank 294
LDS (Church of Jesus Christ of Latter-day Saints) 35
League of Nations 65, 166, 192
Leavenworth, Kansas 76, 98, 99, 100, 102, 103, 104, 105, 106, 107, 110, 119, 127, 135, 139, 286, 290, 307, 313, 314, 315
Leavenworth, Colonel Henry 100
Lee, General Robert E. 150
Lend-Lease Act 205
Leningrad, Russia 279
Leota, Kansas 7
Leta 59, 60, 62, 70
Letters from Mit 299, 302, 309
Lewis, Cornelia Byram 19
Lewis, Second Lieutenant John E. 19
Lexington Intelligencer 119, 125, 137
Lexington Junior-Senior High School, Lexington, Missouri 116, 131, 132, 135, 137
Lexington Masonic Lodge #149 124
Lexington, Missouri 110, 111, 112, 113, 114, 115, 116, 117, 118, 119, 121, 122, 124, 125, 126, 127, 128, 129, 130, 131, 132, 133, 134, 135, 136, 137, 161, 171, 257, 258, 300, 306, 313, 315, 316, 319
Life magazine 205, 215, 318
Lincoln, President Abraham 150
Livesay, General William G. 124, 226
Livorno, Italy (Leghorn) 260
Long Island, New York 47, 48, 49, 311
Lost Battalion 50
Louisiana State Seminary of Learning & Military Academy
Louisiana State University 103
Luftwaffe 134, 261
Luftwaffe Field Division (LFD) 261
Luxembourg 51
Lyon, Brigadier General Nathaniel 113

M

MacArthur, Brigadier General Arthur 145
MacArthur, General Douglas 100, 145 (cadet), 154, 175, 176, 190, 191, 205, 307 (MacArthur Memorial), 318
Macdonald, Dwight 166, 317
Madonna of the Trail 112
Magill, Colonel W. Fulton 259
Mainstreet Theatre 116
Maize, Captain Lester 278
Major, Captain Clarence 279
malaria 5, 206, 300
Manchester, Connecticut 39, 48, 49, 127, 164
Manhattan, Kansas 193, 206, 207, 240, 255, 283
Manhattan Republic 284
Manila Bay 174, 188, 193
March, Shoot and Obey! 219, 244
Marianas 192
Marine Corps (USMC / Marines) 170, 230, 256

Mark IV tank 277
Marseilles, France 58, 232
Marshall, General George C. 92, 213, 313, 314, 319
Marsh, Major Arthur H. 291
Martinville Ridge 266, 267, 268, 269, 270, 271
Maryland 70, 82, 112, 241, 248, 250, 272, 274, 280, 286, 293, 306, 320, 321
Masons (Masonic Lodge) 114, 124, 136, 257
Massachusetts 6, 39, 48, 84
Mayen, Germany 52
McAuliffe, General Anthony 159
McCain, General Henry P. 25, 26, 40, 42
McCain, Senator John 25
McClellan, General George 150
McCully, Andrew "Mac" 140
McCully, Catherine 163
McKee, Corpsman Don 278
McKinley, President William 8
Medford, Oregon 219
medic 262, 277, 279
Mediterranean Sea 231
Merchant Marines 230
Merchantville, New Jersey 83, 88, 89, 95, 140, 163, 292, 307, 313
Meuse-Argonne, France 42, 109, 130
Mexican Punitive Expedition 24
Mexico 22, 24, 292, 293, 294, 310
Milford, Delaware 298
military cities 25, 39, 312
Military Science 82, 83, 110, 114, 115, 124
Millard Preparatory School / Millard, Homer B. and Rose 159, 160
Miller, Reverend Luther D. 107, 108, 314
Miller, Jr., Reverend Luther D. Miller 108, 307
Millman, Morton 201

Miniature Ring 195
Missouri 8, 11, 26, 110, 112, 113, 117, 118, 122, 123, 126, 127, 129, 130, 132, 135, 136, 137, 156, 257, 258, 290, 300, 307, 315, 316, 320
Missouri Compromise 113
Mix, Tom 117, 127
Molotov-Ribbentrop 166
Monte Carlo, Monaco 58
Monte Cassino, Italy 258
Moraine Park School, Dayton, Ohio 198
Moravia, New York 198
Moro 180, 181, 183, 318
Morris, Private First Class William L. 264
Moselle River 53, 57
Mulberry harbors 244
Mulheim, Germany 56
Munich Agreement 166
Murphey, Lieutenant Colonel William W. 301
Murphey, Mae 170, 178, 185, 186, 192, 193, 206, 207, 317
Mussolini, Benito 68, 253

N

Nankivell, Ted 158, 301
Naples, Italy 240
Nash automobile 82, 96, 307
National Cathedral, Washington, DC 208
National Defense Acts 22, 71, 72, 99, 114
National Guard 22, 24, 25, 26, 30, 72, 92, 130, 158, 219, 310, 321
National Socialist Workers Party (NSDAP/Nazi Party) 64, 68
National Zoo (Smithsonian) 179, 192, 298, 306, 318, 320
Navy 122, 123, 139, 152, 170, 184, 188, 190, 192, 204, 209, 230, 256, 312

Netherlands 53
New Deal 101, 118, 133
New Jersey 6, 38, 50, 52, 83, 88, 140, 163, 164, 196, 198, 202, 204, 208, 215, 281, 292
New Rochelle High School, New York 201
New York 9, 48, 49, 50, 76, 85, 106, 116, 135, 139, 140, 152, 157, 160, 162, 163, 164, 167, 169, 191, 198, 200, 201, 202, 203, 213, 254, 273, 295, 306, 307, 311, 313, 316, 317, 319, 321
New York *Post* 202
New Zealand 298
Niland brothers 302
Nininger, Sandy 300
Normandy 232, 240, 243, 245, 246, 249, 252, 261, 265, 266, 274, 277, 279, 281, 290, 302, 307, 320, 321
Normandy American Cemetery 290, 320
North Africa 104, 231, 240, 254, 259, 305
Norton Cemetery 302
Norton, Kansas (Norton County) 5, 6, 7, 8, 9, 10, 11, 12, 13, 14, 15, 18, 20, 23, 27, 28, 29, 30, 37, 39, 49, 74, 81, 111, 128, 129, 293, 302, 306, 309

O

Oak Park, Illinois 200
Officer Training Camps (OTC) 27
Old Fuss and Feathers 75
Olmsted, Frederick Law 76
Omaha Beach 245, 246, 248, 270, 273, 274, 290, 321
Ondrick, Lieutenant Colonel John G. 288
One Hundredth Night Show 155, 158
Operation Avalanche 253

Operation Bodyguard 232
Operation Bolero 241
Operation Cobra 273, 276, 277
Operation Overlord 241
Operation Vendetta 232, 238
Ortlund, Ray 301
Ottoman Turkey 51

P

Pacific Ocean 169, 170, 223, 302
Pacis, Vicente Albano 191
Pact of Paris 100
Palm Springs, California 202
Panama Canal 170
pandas (Hsing-Hsing and Ling-Ling) 298
Panzer Corps 261
Panzer Division 277
Parade Rest 287, 288
Paris, France 52, 57, 58, 100, 273
Patton, General George S. 71, 159, 241, 312, 319
Pearl Harbor 170, 202, 206, 207, 301, 318
Peck, Jack 205
Pendergast, Tom 117
Pendergrast, Captain Alan 106
Pendergrast, Bill 105
Percy, France 276, 277
Pershing, General John J. 19, 24, 52, 65, 77, 103, 311
Philadelphia, Pennsylvania 82, 83, 84, 86, 89, 115, 152, 204, 223, 225, 313
Philippine Department 173, 175
Philippine Islands 105, 136, 137, 154, 161, 164, 169, 170, 174, 175, 176, 178, 179, 180, 188, 190, 192, 193, 205, 206, 213, 214, 294, 300, 301, 318; Luzon 176, 180; Manila 171, 172, 173, 174, 175, 176, 178, 180, 181, 184, 185, 188, 189, 190, 191, 192,

193, 206, 300, 317
Philippine Military Academy 175
Philippine Scouts 175, 300
Photoplay magazine 199
Picatinny Arsenal, New Jersey 38, 48
Pittsburgh, Pennsylvania 198, 199, 200, 291
Playland Park, Rye, New York 203
Plebe 141, 142, 144, 145, 147, 152, 153, 155, 158, 159, 220
Plymouth automobile 96, 215
pneumonia 38, 40, 41, 49, 128, 130
pneumonic plague 41
Polla, Hector 300
POW 300, 301, 318, 319
"Powder River! Let 'er Buck!" 219
Presidio, San Francisco, California 74, 206
Preston, Sergeant Clayton P. 35
Price, Major General Sterling 113
Prohibition 62, 117, 118
Pryor, Helen Sloman 198, 199, 291
Pueblo, Colorado 35, 307
Purnell, Lieutenant Colonel William 248, 266, 275, 281, 286

Q

Quartermaster Corps (QMC) 52, 53, 54, 56, 290
Quezon, Manuel 175

R

RAF 231
Reader's Digest 289
Reed, Arthur 29
Reed, Charles (Charlie) 9
Reed, Elizabeth Crandall 114, 130, 161, 294, 297, 298, 307
Reed, Eva Mae 29, 95
Reed, Laura Sloman (*See* Vaughan, Laura Sloman Reed)
Reed, Mark 298, 305
Reed, Ollie William 3, 6, 7, 8, 9, 10, 12, 13, 14, 15, 16, 19, 20, 21, 22, 23, 24, 25, 26, 27, 28, 29, 30, 31, 32, 33, 35, 36, 37, 38, 39, 40, 42, 47, 48, 49, 50, 52, 53, 54, 55, 56, 57, 58, 59, 60, 61, 63, 66, 67, 69, 70, 72, 73, 74, 78, 80, 81, 82, 83, 86, 87, 88, 89, 91, 92, 93, 94, 95, 96, 98, 99, 100, 102, 104, 105, 106, 108, 109, 110, 111, 114, 115, 118, 120, 121, 122, 123, 124, 125, 128, 129, 134, 135, 136, 139, 140, 141, 144, 148, 161, 163, 164, 168, 169, 172, 174, 176, 177, 178, 179, 180, 186, 187, 189, 193, 205, 206, 207, 212, 214, 221, 222, 224, 225, 227, 232, 233, 234, 235, 238, 239, 240, 241, 243, 244, 245, 246, 248, 249, 250, 251, 252, 255, 256, 267, 269, 272, 273, 275, 279, 281, 283, 284, 285, 288, 289, 290, 291, 292, 295, 296, 298, 300, 302, 305, 313, 319
Reed, Jr., Ollie William (Buddy/Bud) 3, 43, 44, 45, 48, 49, 54, 55, 56, 59, 60, 61, 62, 66, 70, 73, 74, 81, 91, 94, 97, 98, 105, 106, 107, 109, 110, 111, 119, 120, 121, 122, 123, 127, 128, 129, 130, 131, 132, 133, 134, 135, 136, 137, 139, 140, 141, 143, 145, 146, 147, 148, 152, 153, 154, 155, 157, 158, 159, 160, 161, 162, 164, 169, 172, 178, 179, 185, 195, 196, 197, 203, 204, 205, 207, 208, 209, 210, 211, 212, 214, 215, 216, 217, 218, 219, 220, 221, 222, 223, 225, 226, 227, 229, 230, 231, 234, 237, 238, 239, 240, 242, 243, 244, 246, 252, 253, 254, 255, 256, 257, 258, 261, 263, 264, 273, 284, 285, 290, 291, 295,

296, 300, 301, 315, 319, 320
Reed, Mabel 163, 216
Reed, Maryalyce (*See* Jenkins, Maryalyce Reed)
Reed, Mary Plusky 6, 129
Reed, Mildred Boddy 3, 5, 6, 7, 10, 11, 12, 14, 15, 16, 17, 18, 19, 20, 22, 23, 27, 28, 29, 30, 32, 35, 36, 37, 38, 39, 42, 47, 48, 49, 50, 54, 55, 56, 57, 58, 59, 60, 61, 66, 70, 73, 74, 81, 82, 86, 87, 88, 89, 91, 94, 95, 96, 97, 98, 106, 107, 108, 123, 125, 126, 127, 128, 129, 132, 133, 134, 135, 137, 140, 141, 145, 153, 156, 161, 162, 163, 164, 168, 169, 170, 171, 172, 176, 177, 178, 179, 180, 181, 184, 185, 186, 187, 188, 189, 192, 193, 206, 207, 209, 211, 212, 213, 214, 215, 221, 222, 225, 227, 234, 235, 238, 239, 240, 241, 242, 243, 245, 246, 251, 255, 257, 273, 275, 283, 284, 285, 286, 288, 289, 290, 291, 292, 293, 294, 295, 296, 299, 300, 301, 302, 303, 305, 309, 313, 321
Reed, III Ollie William (*See also* Vaughan, Bill) 222, 224, 225, 227, 239, 240, 243, 246, 255, 256, 273, 283, 291, 292, 295, 298, 319
Reed, Orville 6, 7, 129
Reed, O.W. 14, 38, 137, 288
Reed, Theodore Harold "Ted" 70, 73, 74, 81, 83, 91, 94, 97, 98, 105, 106, 107, 108, 111, 119, 120, 125, 126, 127, 128, 135, 136, 137, 158, 161, 163, 164, 165, 167, 168, 172, 179, 180, 181, 185, 186, 192, 193, 206, 207, 216, 218, 234, 235, 239, 240, 243, 251, 256, 273, 284, 285,

290, 294, 295, 297, 298, 301, 302
Regular Army 72, 73, 82, 92, 101, 124, 248, 314
Reparations Commission (Reparations) 64, 67
Reserve Officer Corps 85
Reserve Officer Training Corps (ROTC) 21, 82, 85, 86, 87, 110, 115, 124, 135, 136, 309
Reserves 22, 72, 92
Retzer, Karl 160
revolution 51, 67
Rheinische 63
Rhineland, Germany 51, 52, 53, 60, 63, 64, 65, 67
Rhine River 52, 53, 60, 61, 65, 281
Rhode Island 50
Rice, General John K. 287
Riparbella, Italy 261
R.M.S. *Lusitania* 21
Rockenbach, Colonel Samuel D. 70
Rome, Italy 58, 253, 254, 258, 273, 320
Rooney, Mickey 275
Roosevelt 9, 118,
Roosevelt, President Franklin Delano 133, 134, 137, 161, 176, 198, 202, 205, 317
Roosevelt, Teddy 9
Root, Elihu 85
Rotterdam, Netherlands 53
Russia 32, 51, 64, 166, 191
Russian czar 51

S

S-3 Plans and Training Officer 173, 176, 205
Sainte-Croix Church, Saint-Lô 273
Saint-Lô, France 245, 248, 266, 267, 268, 269, 270, 271, 272, 273, 274, 276, 277, 321
Saint-Lô-Bayeux highway 266
Salida, Colorado 34, 310
Salt Lake City, Utah 30, 31, 32, 35, 310

San Francisco, California 170, 178, 192, 193, 206, 317
Santa Fe Trail 100, 113
Sarvis, Betty 29, 49, 127, 128, 129, 131, 132, 295, 296
Sarvis, Christine 29, 39, 49, 127
Sarvis, Joe 28, 39, 127
Sarvis, Lucinda Biggs 6, 39
Sarvis, Mae 29
Sarvis, Myra 15, 294
Sarvis, Nell 29
Saturday Evening Post 203
Saving Private Ryan 302
Scaife, Christopher T. 200
Scaife, Jane Katherine 200. (*See* Sloman, Catherine)
Scarsdale, New York 201, 306, 321
Schantz, Adam 198
School of Application for Infantry and Cavalry 99, 100
scouting / Boy Scouts 125, 126, 316
Sebastian, Lieutenant Colonel Henry G. 192
Sedgwick County Zoo, Wichita, Kansas 298
Selective Service Act 25, 207
Sellers family 114, 257
Sellers, Colonel James M. 115, 124, 135
Sellers, Sandford 114, 115
Sellers, William Wentworth 114
Seventh Corps Area 124
Severn River (Maryland and UK) 70, 73, 241
Shanghai Bowl 191
Shanghai International Settlement 190, 191
Sheen, Lieutenant Colonel Henry H. 56
Shenandoah National Park 217
Sherman, General William Tecumseh 103, 150
Sherman, Sergeant Abe 280
Sherwood Forest, Maryland 70
Silver Star 278, 280, 318, 319
Simpson, Lieutenant General William H. 242
Sino-Japanese War 134, 191
Skorzeny, Otto 253
Skyline Drive 217
Slingluff, Captain Kernan 246
Sloman, Blanche Griffith 197
Sloman, Catherine/Katherine Kabel Scaife 197, 200, 201, 202, 203, 210, 216
Sloman, Christopher 197, 200, 201, 202
Sloman, James 84, 115, 124, 150, 174, 197, 198, 199, 201, 202, 277, 296, 307, 315, 317
Sloman, Laura (*See* Vaughan, Laura Sloman Reed)
Sloman, Laura Gillmore 198, 199, 200
Sloman, Mabel 201
Sloman, Mabel Jane (*Also see* Gillis, Ann) 201
Sloman-MacKinnon 201
Sloman, Margaret 107, 196, 197, 200, 201, 202, 203, 216, 306
Sloman, Morris/Michael/Mike 197, 202, 204, 209, 210, 255, 315
Sloman, Richard 197, 201
Smith, Page 160
SMS Cormoran 31
Sole Survivor Policy 302
Spanish-American War 34, 101, 170, 175
Spanish Civil War 134
SS *Athenia* 170
SS *President Coolidge* 206
SS *William B. Giles* 229
SS-Panzer 262
Stanley, "Red" 202
State, War, and Navy Building 139
Stayer, Lieutenant Colonel Morrison C. 92

Steuben, Baron Friedrich Wilhelm von 77
Stillwell, Joseph W. "Vinegar Joe" 92
St. Laurent-sur-Mer, France 290
St. Mary's Hall 204
Star Spangled Banner 210, 275
Stourport-on-Severn, West Midlands 241
Strength for Service to God and Country 246, 264, 285, 319
Strickland, Newton Harrell 164
Strickland, Sonny 98
Stroh, Captain Donald A. 105
Strong, Robert 160
Student Army Training Corps (SATC) 19, 21, 22, 114
Subic Bay 189, 190
Sullivan brothers 302
Sumay, Guam 171, 172
Summers, Lieutenant Odis B. 280
Sweet family 32, 98, 106, 296, 301
Sweet, Ann 98
Sweet, Betty 296
Sweet, Captain Robert 98, 301
Sweet, Brigadier General Joseph B. 98

T

Tagalog 184
Taggart, Reverend Gerard 278, 279, 321
Tank School 70, 71, 72, 312
Taps 121, 275
television 167
Temple, Shirley 202
Terry, Lieutenant Colonel William T. 274
Tessy, France 250, 277
Tharp, Patricia 12, 16, 20, 305, 309, 321
Thayer, Sylvanus 150, 154
Theodore Roosevelt High School 202
Tiber River 260
Townsend, Edith 193
Townsend, Major Glen R. 192
Treaty of Versailles 52
Truman, Harry S. 112
Turner, Staff Sergeant Arlyn R. 264
Two-Nine, Let's Go! / 29, Let's Go! 244, 271, 275, 281
Tyrrhenian Sea 260

U

U-boat 170
UGS-40 convoy 229, 231
Ulio, Major General James A. 284
United States Military Academy (USMA/West Point) 19, 48, 52, 101, 103, 104, 109, 110, 120, 124, 133, 134, 135, 136, 137, 139, 140, 141, 142, 144, 145, 146, 148, 149, 150, 151, 152, 153, 154, 155, 156, 158, 159, 160, 161, 162, 164, 172, 175, 178, 179, 195, 196, 197, 203, 204, 205, 207, 208, 210, 212, 213, 214, 215, 216, 220, 223, 234, 246, 247, 264, 291, 299, 300, 301, 306, 316, 317, 318, 319
U.S. Army 26, 55, 100, 170, 173, 175, 213, 216, 286, 288, 312, 313, 314, 318, 319, 320
  Armies
    First Army 252
    Fifth Army 253
    Third Army 52, 241, 242
  Corps
    XII Corps 234, 241, 242
    XIX Corps 248, 271
  Divisions
    1st Armored Division 104
    2nd Armored Division 276
    3rd Armored Division 319
    3rd Infantry Division 206
    12th Infantry Division 40, 42
    29th Infantry Division 109, 244, 245, 248, 249, 250, 252,

261, 269, 270, 271, 272, 274, 275, 276, 277, 281, 285, 305, 306, 320, 321; 29ers 270, 272, 276, 277
30th Infantry Division 206, 207, 248
32nd Division 92
34th Infantry Division 258, 259
35th Infantry Division 130, 249
76th Division 40
77th Infantry Division 49, 50, 311
78th Infantry Division 212, 221, 225, 232, 287; Lightning Division 288
88th Division 37
89th Infantry Division 109
91st Infantry Division 109, 124, 219, 221, 223, 225, 226, 232, 237, 238, 240, 243, 244, 253, 258, 261, 264, 305, 306, 319, 320, 321
104th Infantry Division 221

Regiments
10th Cavalry Regiment 19
29th Infantry Regiment 91, 93, 94, 96, 141, 214, 246
31st Infantry Regiment 171, 173, 174
42nd Infantry Regiment 30, 31, 32, 33, 34, 35, 37, 39, 40, 47, 48, 50
50th Infantry Regiment 50, 52, 313
115th Infantry Regiment 270
138th Infantry Regiment 130
175th Infantry Regiment 244, 246, 248, 250, 266, 267, 268, 269, 272, 274, 277, 278, 279, 280, 281, 321
309th Infantry Regiment 232, 233, 234, 235, 249, 287, 288, 289, 300

361st Infantry Regiment 258
363rd Infantry Regiment 226, 231, 240, 259, 260, 261, 263, 264, 320

Battalions
316th Medical Battalion 262
747th Tank Battalion 267, 269

Artillery Batteries
110th Field Artillery 251
224th Field Artillery 280

U.S. Army Transport (USAT) 54
USAT *U.S. Grant* 169, 170, 171
USAT *Buford* 54, 55
USAT *Cambrai* 65, 66
USAT *Republic* 193 51, 64, 67, 193, 284

USO 276
USS *Arizona* 301
USS *Madawaska* 170
Utah 28, 30, 31, 35, 48, 52, 302, 310

V

Vada, Italy 264
Valle de Bravo, Mexico 293
Vallee, Rudy 201, 202
Van Vliet, Colonel John H. 234
Vaughan, Laura Sloman Reed 196, 197, 199, 201, 203, 204, 205, 208, 209, 210, 211, 212, 215, 216, 217, 218, 220, 221, 222, 223, 224, 225, 226, 229, 230, 234, 235, 237, 238, 239, 240, 243, 246, 254, 255, 256, 257, 258, 261, 263, 264, 273, 283, 284, 285, 289, 290, 291, 295, 318
Vaughan, Lieutenant William 291, 295
Vaughan, Bill (See also Reed, III, Ollie William) 295, 298, 299, 305, 319
Velie, Lenora 216, 306
Venice 57, 58, 59
Vessie, France 245
veterinary medicine 179, 240, 251

Victory-mail (V-Mail) 237, 273
Vietnam (French Indochina) 192
Villa, Pancho 20, 24
Villebaudon, France 277, 278, 279, 304
Vire River, Vire, France 247, 276
Vixen Tor 245
Vogt, Julia Ann 297
von Steuben, Baron Friedrich Wilhelm 77

W

Wake Island 190, 317
Walter Reed Army Hospital 70
Walter Reed General Hospital 160
War College 100, 102, 103, 109, 110, 140, 306
War Department 65, 84, 93, 109, 114, 124, 219, 222, 285, 310, 320, 321
Ward, Lealon 294
Ward, Myra Sarvis 294
War of 1812 78, 100, 150
War Plan ORANGE 190
War Plan RAINBOW 190
War Plans Division 100
Washington, DC 76, 108, 139, 159, 163, 174, 207, 294, 297, 299, 305, 312, 314, 318, 319, 320, 321
Washington, President George 148
Week End Pointer 196
Wehrmacht 168, 277
Wells, Brigadier General Briant H. 76
Wells, Thomas 141
Wentworth Military Academy 110, 114, 115, 116, 119, 120, 212, 122, 123, 124, 127, 130, 133, 134, 135, 136, 158, 257, 300, 301, 306, 315
Westchester County, New York 201
Western Union 284
Westinghouse 168
West Midlands, UK 241
West Point (*See* United States Military Academy)
West Shore Line 196
What's the Matter With Kansas? 8
"When it's Apple Blossom Time in Normandy" 265
Whistler, James Abbot McNeill 150
Whitehouse, Musician Third Class Fred T. 35
White, Major General George 219
white tiger 298
Wichita, Kansas 298
Wikoff, Lester B. 115
William Jewell College 135, 156, 195, 296, 307
William Jewell *Student* 156
Williams, Reverend Roy 216
Wilson, President Woodrow 9, 14, 21, 22, 24, 65, 69, 77, 108, 278, 313, 321
Women's Air Corps (WACs) 256
Wood, General Leonard 85, 309
Woodring, Harry 124
Woods, Ralph 141
World of Tomorrow 163, 164, 165, 167, 317
World's Columbian Exposition 6, 76, 164, 165
World's Fair (World of Tomorrow) 163, 164, 166, 168, 202, 317
World War I 20, 40, 42, 51, 64, 69, 70, 92, 100, 106, 108, 115, 197, 249, 272, 280, 317, 318 (See also First World War and The Great War); War to end all wars 69, 108
World War II 51, 80, 92, 104, 106, 191, 214, 241, 242, 253, 259, 264, 281, 299, 300, 301, 305, 315, 316, 318, 319, 321
Wright, Frank Lloyd 200
Wyre Forest, UK 241, 319

Y

Yaphank, Long Island, New York 48
Yearling 142, 161
Yingling, C.E. 130
Yingling, Mattie 130, 131
Yonkers High School 202
Young, William 198

Z

Zamboanga, Philippines 180, 181
Zoology 298

About the Author

Dennis Whitehead is a writer, photographer, and producer living in Arlington, Virginia. His previous book, *The Day Before the War*, recounts the events along the German-Polish border on August 31, 1939 that ignited World War II in Europe.

He has a lovely wife, two grown children, and two dogs.

www.loveandsacrifice.com
www.dennis-whitehead.com

www.ingramcontent.com/pod-product-compliance
Lightning Source LLC
LaVergne TN
LVHW051541070426
835507LV00021B/2366